WHAT THE EAR HEARS
(AND DOESN'T)

Inside the Extraordinary Everyday World of Frequency

RICHARD MAINWARING

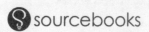

Published by Sourcebooks
P.O. Box 4410, Naperville, Illinois 60567–4410
(630) 961-3900
sourcebooks.com
Originally published in 2022 in Great Britain by Profile Books

Library of Congress Cataloging-in-Publication Data

Names: Mainwaring, Richard, author.
Title: What the ear hears (and doesn't) : inside the extraordinary everyday world of frequency / Richard Mainwaring.
Description: Naperville, Illinois : Sourcebooks, [2022] | Includes bibliographical references and index. | Summary: "In 2011, without warning, a skyscraper in South Korea began to shake uncontrollably and was immediately evacuated. Was it an earthquake? A terrorist attack? No one seemed quite sure. The actual cause emerged later: Twenty-three middle-aged Koreans were having a Tae Bo fitness class in the office gym on the twelfth floor. Their beats had inadvertently matched the building's natural frequency, and this coincidence caused the building to shake at an alarming rate for ten minutes. Frequency is all around us, but really isn't understood. What the Ear Hears (and Doesn't) reveals the extraordinary world of frequency-from medicine to religion to the environment to the paranormal-not through abstract theory, but through a selection of small memorable human (and animal) stories laced with dry humor, including: The elephant who anticipated the 2004 Asian tsunami and carried a young girl miles inland, saving her life The reason for those deep spiritual feelings we have in churches The cutting-edge methods that are changing medicine The world's loneliest whale"-- Provided by publisher.
Identifiers: LCCN 2022026566 (print) | LCCN 2022026567 (ebook) | (trade paperback) | (epub) | (adobe pdf)
Subjects: LCSH: Sound-waves--Miscellanea | Sound--Miscellanea | Electromagnetic waves--Miscellanea,
Classification: LCC QC229 .M35 2022 (print) | LCC QC229 (ebook) | DDC 531/.1133--dc23/eng/20220622
LC record available at https://lccn.loc.gov/2022026566
LC ebook record available at https://lccn.loc.gov/2022026567

Printed and bound in Canada.
MBP 10 9 8 7 6 5 4 3 2 1

For Rose and Jack
Always find fascination in
the world around you xx

THE KEY

A♭—Prelude 1

A—The Black Hole at the End of My Piano 13

B♭—Pitch Perfect 32

B—Bridges over Troubled Waters 52

C—Ghost Notes 77

D♭—Fix in the Mix 98

D—Only the Lonely 121

E♭—Of Sound Mind 147

E—Somewhere over the Rainbow 162

F—Acoustic Ammunition 186

G♭—Sleight of Ear 213

G—High Times 237

Sources 267
Musical Frequency Table 293
Acknowledgments 297
Index 301
About the Author 311

A **FINITE SECTION** *of*

0.0000000000000003 Hz
Lowest note in the universe 5 pages to the left!!

4 Hz
Camster Round
neolithic grave

18.98 Hz
Tandy's
ghost

40 Hz
"Another One
Bites the Dust"
bass guitar

0.001 Hz
tsunamis are on
the next page

16.35 Hz
Franck's
organ pipe C

20 Hz
edge of
human
hearing

27.5 Hz
lowest note on
standard piano

the INFINITE PIANO

10000000000000000000000 Hz

Gamma rays about 6 pages to the right!!

50 Hz
Szechuan pepper 'tingle'

82.4 Hz
Dick Dale's "Misirlou"

440 Hz
oboe's concert A

1000 Hz
Allegri's *Misere* high vocal C

52 Hz
loneliest whale's call

250 Hz
honey bees' buzz

750 Hz
Eine Kleine Nachtmusik first violin

2300 Hz
Vienna city bird calls

PRELUDE

In the days leading up to November 14, 1940, British intelligence mixed up two musical notes (the two found at the start of the vocal line of the swing song "La Mer"). The consequence was that, on the morning of November 15, the English city of Coventry suffered one of the worst bombing raids of the Second World War. It is impossible to say how many lives could have been saved had the notes G and C been correctly identified.

Allow me to explain.

In the deadly skies of 1940, the German military were technically far superior to the Allies, and one of their greatest secret weapons was their inspired radio guidance technology. Thanks to the Lorenz system invented in Berlin in the early 1930s, aircraft could land guided by precise radio beams, dispensing with the need for visual contact with a runway. With some minor modifications, the Luftwaffe initiated the Knickebein project to assist them during air raids, ingeniously adapting the Lorenz system, which guided their bombers along two almost parallel radio beams, one of dots to the left and the other of dashes to the right. If a pilot could fly perfectly down the bowling lane between them, he would hear a mix of these dots and dashes as a single continuous tone, directing him toward a specified bombing target. Drift to the left and the continuous tone was replaced by dots; drift to the right and he would hear only dashes. Thanks to Knickebein, neither clear conditions nor daylight were essential for successful bombing raids. The beams were transmitted from Nordfriesland near Germany's border with Denmark, Kleve near the Dutch border, and Lörrach in the southwest. But in 1940, very few in the British military believed the Germans had the technology to send highly accurate radio beams from continental Europe over English skies.

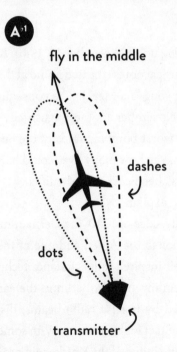

The Lorenz Principle

A twenty-eight-year-old PhD science graduate from Oxford begged to differ. In 1939, Reginald Victor Jones was the first scientist ever assigned to British intelligence. In 1940, he found himself—as assistant director of Intelligence (Science)—bending the ear of Winston Churchill at 10 Downing Street. He told the prime minister that the Germans had constructed aerials he suspected could broadcast two radio signals from the same location, remaining only yards apart at their ultimate target in England— like a very narrow flashlight beam. This would allow German bombers to navigate along such beams, dropping their lethal cargos with frightening precision.

Churchill had a hunch that R. V. Jones was something special. Undeterred by the skepticism of his military bigwigs, he gave the young man permission to put his theory to the test. An aircraft searched the sky, listening for the theoretical beams, and eventually, after some early failures, pilot Flight Lieutenant Bufton and his observer Corporal Mackie found them. Jones was informed that the Germans were broadcasting dots and dashes at 1,500 hertz (hertz is the internationally recognized unit of frequency, often abbreviated to Hz). This is the musical note G_6, found at the end of Tomaso Albinoni's Adagio in G Minor—I'll explain what G_6 means shortly.

Jones now had a dilemma. If the Royal Air Force and ground defenses ambushed and destroyed the German bombers flying in formation along the Knickebein line, the Nazis would know that the British were on to them. It seemed like there was no choice but to let the Germans fly on undisturbed. Jones opted for a more wily approach, broadcasting a rival British G_6 note of 1,500 Hz, one that intersected the Germans' signal but aimed away from the Luftwaffe's intended target. If the bombers could be fooled into following the British G_6 tone, they could be guided to drop their loads in much less densely populated or strategically sensitive areas.

Jones was justifiably very pleased with himself. He had proved that the Germans' Knickebein project was real, that they were broadcasting radio beams across English skies, and that there was potential to blunt the Luftwaffe's attack via a 1,500 Hz counter tone. Using information gleaned from the code-breaking Enigma machine, British intelligence predicted that nighttime bombing raids were soon to target cities in the Midlands, particularly Coventry and Birmingham. Churchill's bigwigs had to stand aside while Jones, the young genius of British intelligence, led the innovative newfangled defense against the might of the German Luftwaffe. But all was not well; the bombs still fell on Coventry.

Jones wrote about the discovery of the two-tone mix-up in his spellbinding 1978 book *Most Secret War*: "the Royal Aircraft Establishment at Farnborough found that the filter [in a shot-down German bomber] was tuned to two thousand cycles per second, a high-pitched note corresponding roughly to the top 'C' on a piano. Our jammers had been set not on this note but on one of fifteen hundred cycles per second, corresponding to the note 'G' below top 'C.'"

The Germans' note was actually a C at 2,000 Hz and not a G at 1,500 Hz. Indeed, their radios had even been set to filter out any notes beneath 2,000 Hz. This meant that the British had been trying to fool the Germans with the wrong note; the bombers' radios never even heard the British counter beam.

In his book, Jones used a rather inventive musical means of explaining his story, illustrating the frequencies of the two radio signals with corresponding notes on a piano. Understanding that the British were 500 Hz away from the "correct" frequency is of interest, though a little meaningless to the reader. However, knowing that the mistaken pitch was a mere three piano notes away from success gives us a clearer sense of why Jones was so exasperated by the failure. "Of all the measurements in connection with the German beams, easily the simplest was to determine the modulation note [the 1,500 Hz tone]," he wrote, "and yet whoever had done it had either been tone deaf or completely careless... I was so indignant that I said that whoever had made such an error ought to have been shot."

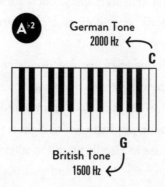

It seems that Jones's brilliant mind knew no bounds. As well as spotting cunning Nazi technological innovations, Jones the author had an ingenious way of simplifying the complexities of frequency, combining the often-meaningless unit of hertz with the much more familiar concepts of musical pitch and a piano keyboard. In homage to Jones's creative and illuminating trope, *What the Ear Hears (and Doesn't)* extends the concept, exploring *all* frequencies within a musical context, using the familiarity of a piano keyboard to help us understand, hear, and marvel at the extraordinary world of vibration, waves, and frequency.

Welcome to the Infinite Piano

Here's a quick test for you: sing the note of the frequency 123 Hz. I assume you're struggling. It's probably easier if I reveal that

123.47 Hz is the first guitar note of the Rolling Stones' "(I Can't Get No) Satisfaction." How about 392 Hz—can you hum that? Again, it's far more achievable if you know that 391.99 Hz is the pitch of the initial three Gs played by the violins at the start of Ludwig van Beethoven's Symphony no. 5. When presented with musical examples, your memory can give you an approximate equivalent tone for that frequency. Such aural signposts feature throughout this book, providing musical reference points for our fascinating journey.

However, frequencies are not exclusive to the range of a piano or the spectrum within which music lies. Above the highest piano C at 4,186 Hz (this is C_8, numbered so because it's the eighth C you'll encounter from left to right on a standard piano), frequency continues to rise through the ultrasonic range as well as through microwaves, X-rays, and gamma rays. And beyond the lowest bass note, measurable vibrations descend forever. But what if a piano could extend its range to cover all these frequencies? If the ivories continued beyond that top C, where would those imaginary notes lead to? As light is a wave with a set of frequencies, how far along a piano keyboard would one have to travel to "play" a rainbow? And what would it sound like? Similarly, how many octaves to the left of a piano's bottom A at 27.5 Hz (frustratingly not numbered A_1 but A_0) would one have to plunge before one could find the notes of tsunamis and earthquakes? Welcome to my new invention—the Infinite Piano.

As a space-obsessed child (I persuaded my parents to let me skip school to watch the first space shuttle launch live on TV in 1981), I dreamed of donning a tinfoil suit and exploring the mysteries of the universe. When I lowered my gaze (and ambition) and decided to become a musician instead, little did I know that the cosmos would one day come to me. Could it be that all the violin vibrations that first resonated through me as a five-year-old and that have continued throughout my professional life inspired my interest in frequency? Or perhaps my compositional exploration of new and extreme sounds and pitches sparked a desire to search beyond the limited horizon of a "normal" piano's range? Whatever it may be, my study of the narrow spectrum of mechanical vibration that is music has led me to create an instrument that puts the universe at my fingertips, dispensing with the need for an external oxygen supply or a billion-dollar spaceship. Call it compensation for the fact that I never got to ride a rocket or even see that first space shuttle launch—it was delayed for a few days, and there was no chance of getting a second day off school.

Before we experience a piano scale like no other, though, it might be best to clear up some important questions. First, what are "hertz"? Frequency used to be measured in cycles per second, a cycle being one complete oscillation of a wave; if the duration of this was one second, the old nomenclature would call that one cycle per second (1 cps). In 1935, the International Electrotechnical Commission renamed this measurement in honor of the man who proved the theories of electromagnetism, Heinrich Hertz. And in 1960, hertz (Hz) was adopted in the International System of Units as an official unit of frequency; one cycle per second was now 1 Hz.

Second, what is a wave? Physics textbooks will tell you that it is the transmission of information or energy from one place to another without any material object making that journey. I prefer to explain a wave using the classic kids' game "telephone," where a whispered message is passed along a line of children. As with a wave, information is transmitted without anyone moving. However, it's not much of a game if the first child walks to the last child and whispers the message, and it's no longer an example of a wave either. This book focuses on the two main types of wave: mechanical and electromagnetic.

The sound of the German dots and dashes were made through *mechanical* waves, regular changes of air pressure generated by the pilots' headphones, in turn vibrating their eardrums 2,000 times per second (2,000 Hz), the note "top C." Light is also a wave but not a mechanical one—it does not rely on air pressure to travel. It is an *electromagnetic* wave, propagated via atoms and the space in between them; it can travel through a vacuum. For example, red light is a form of electromagnetic wave with a frequency around 430,000,000,000,000 Hz. This means that the wave is vibrating 430 trillion times per second, faster than most of us can comprehend.

But even the awesome frequency of such a complex, non-sounding wave holds no challenge for my Infinite Piano. It is an ingenious and nondiscriminatory instrument. It cares not whether a wave is mechanical or electromagnetic, whether it is audible or silent—it can play all types of waves at all frequencies. But can it truly be called an *infinite* piano? Is there no limit to vibration? Certainly, as it is possible to halve a hertz number forever, we are good to go down at the bass end of our piano. Many eminent physicists claim there is no theoretical upper limit to frequency either, and who am I to disagree? So perhaps my piano really is infinite.

Everyday Vibes—From Cats to Bananas

The impact of frequencies on us humans is staggering. We're shaken by mechanical waves, enabling us to hear and feel. Our brain waves generate frequencies that change throughout our waking day and our sleep. We see vibrations in the form of light, and we crave the frequencies of radiation emitted by fire and the sun that give us warmth. Different parts of our bodies have different resonant frequencies, enabling us to see ghosts or forcing us to hyperventilate during blockbuster movies. And we rightly try to avoid the ultrahigh frequencies of radiation that can severely damage our bodies—unless we need an X-ray or cancer treatment, in which case those same frequencies can be incredibly beneficial. It is also fascinating to explore the impact of frequency on everything else on our planet and even beyond. All living things, from rats to elephants, rely on vibrations. To eat or be eaten is often a matter of a single frequency—if your predator's hearing is finely tuned to 256 Hz and your wings beat at that same frequency, there might be trouble ahead. From communicating where the next meal is to drumming out a courtship dance, from plucking the strings of a finely spun silk web to navigation via the micro vibrations of the mountains, living things rely on hertz.

Cats, as it happens, are the perfect living subjects for this preface. Studies have shown that they purr at 25 Hz and 50 Hz—the musical pitch of A♭. I was initially a little skeptical of this, until I listened to my own cat…and yes, Cysgu (pronounced CUS-kee, the Welsh word for "to sleep") purrs at around 50 Hz. A couple of questions that have perplexed scientists are how and why cats purr. The latest theories suggest, as most of us would have probably guessed, that purrs emanate from a cat's larynx. But *why* they purr is a much more difficult question to answer. Though a purr seems to exude an infectious state of calm and contentment, some cats also purr when they are stressed or anxious. And some purr more often than others—some when they want food, some even when they are eating. But researchers are generally in agreement that the purr of cats (and not only the domesticated type) is centered around 25 Hz or 50 Hz.

Research suggests that the purr is a self-healing device, promoting the growth of healthy bones and tissue. A recent study concluded that "low frequency (25–50 Hz) vibration *in vivo* can... repair bone injury," and "specifically, bones from the 25 Hz and 50 Hz treatment groups displayed increased callus formation." In other words, the notes of Ab_0 and Ab_1 increased the speed at which fractures heal. The frequency of Ab_1, 50 Hz, is the second note of the bass line in Queen's "Another One Bites the Dust" (on the word "bites"). The next time your cat purrs an Ab, consider this—she could be telling you she's content, stressed, or hungry. Or she might want you to leave her alone as she's in the process of repairing her bones. Cats are so enigmatic.

Though I have promised a journey across the earth and out into the universe, there are plenty of frequencies to explore at home. Look no further than your fruit bowl. Bananas and Brazil nuts emit frequencies in the electromagnetic spectrum that we would recognize as radiation. Yes, bananas are radioactive. Indeed, there is an informal unit of measurement of radiation known as the "banana equivalent dose" (BED). Bananas contain a radionuclide, potassium-40, which emits radiation at a frequency between 3,143,400,000,000,000,000,000,000 Hz and 3,530,300,000,000,000,000,000,000 Hz. (To play the lowest frequency of potassium-40 on the Infinite Piano, you would have to walk just under twelve meters or seventy-three octaves to the right of middle C—not that "middle" could exist on an Infinite Piano.) Such radiation, in large doses, plays havoc with the atoms, molecules, and cells in our bodies, causing uncontrolled and random cell division, which can ultimately lead to cancers. However, it has been estimated that one would have to eat 10,000,000 potassium-rich bananas in just one sitting to do any serious radioactive harm—not perhaps the most appealing lunch.

The chapters of this book follow the notes of a piano keyboard. The white note chapters sound a rising scale through the world of frequency, whereas the black notes explore the subtleties of sound and music, areas that have dominated my professional life.

From the ghostly infrasonic vibrations that smear our vision, through the rumbling 52 Hz call of the world's loneliest whale, to the atomic oscillations of lethal "Bikini snow," there is so much to learn and understand about the world of frequency. Come, sit on my Infinite Piano stool, and fasten your seat belt as we embark on an amazing ride through vibration and frequency, from the lowest "note" in the universe through to the astoundingly high frequencies that measure time itself. Our first stop: a black hole.

THE BLACK HOLE AT THE END OF MY PIANO

From the lowest "note" in the universe to the winds of Mars

Sometime between 250 and 300 million years ago, a very low note boomed out across the cosmos. And 250 to 300 million years later, in 2003, a telescope orbiting a tiny speck of rock called Earth detected the note's shock waves rippling through clouds of hot gas. The location was a supermassive black hole in the Perseus galaxy cluster, the telescope was NASA's Chandra X-ray Observatory, and the note was a B-flat.

Chandra's discovery would have made Pythagoras—the Greek philosopher and all-round genius—ecstatic if he had been alive today. Finally, two and a half thousand years after his death, his hunch that celestial bodies have a strong connection with music was confirmed by this telescope. Pythagoras was probably the first person to link patterns in the movements of planets with music. He called this hypothetical celestial music the "harmony of the spheres," and one of its greatest devotees was the seventeenth-century stargazer Johannes Kepler, who constructed a set of mathematical principles to prove Pythagoras right.

In his 1619 book *Harmonices mundi* (*Harmony of the World*), Kepler expanded on the principle that the six then-known planets in the solar system had close musical and mathematical connections. He compared the ratios between each planet's fastest and slowest orbital speeds with the ratios of common musical intervals. An interval is the distance between two musical notes: for example, the first musical leap at the start of "Taps" (the trumpet call performed at Memorial Day and military funerals) is an interval of a fourth, as the leap from G to C contains four steps—G, A, B, C (in our musical alphabet, a new A follows G).

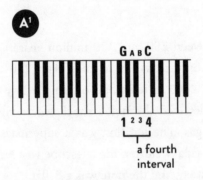

After a series of detailed astronomical observations, Kepler reaffirmed Pythagoras's view that there was a relationship between the planets' orbital velocities and musical intervals. Some of Kepler's mathematical calculations were surprisingly accurate. According to NASA, Earth's minimum and maximum orbital velocities are 29.29 and 30.29 kilometers per second—so the ratio Kepler came up with of 15:16 is impressively close. He also noted that a ratio can be formulated from two frequencies that make a musical interval. In this case, 15:16 is that of a semitone—G_6 is around 1,568 Hz, while a semitone above is approximately 1,661 Hz. (G_6 is the extraordinary whistle tone note

Mariah Carey sings around a minute into "Emotions.") Kepler's argument throughout *Harmonices Mundi* was that such theories, supported by mathematical evidence, proved the wisdom and power of God.

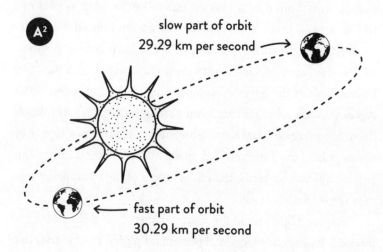

His full ratio table was as follows:

Mercury 5:12 and a minor tenth	There are many minor tenths in the guitar part at the start of Justin Bieber's "Love Yourself." It's a wide leap on the piano, about the span of a large hand.
Venus 24:25 and a diesis	An interval smaller than a semitone; the distance between these two notes makes them sound very slightly out of tune with each other.
Earth 15:16 and a semitone	The first two notes of "The Pink Panther Theme."
Mars 2:3 and a perfect fifth	The opening two notes of the bugle call "The Last Post," or the two that make up the classic power chord found in heavy rock music.
Jupiter 5:6 and a minor third	The distance between the first two pitches of "Greensleeves."
Saturn 4:5 and a major third	The opening leap at the start of the Christmas carol "Once in Royal David's City," or the classic *bing-bong* of a doorbell.

In 2015, I used Kepler's calculations as the basis for a composition that I was commissioned to write for the BBC and the International Space Station. As a presenter of music films on the BBC's prime-time magazine show *The One Show*, my trademark ending would often be a musical stunt—performing as part of a trio in a very small hotel elevator for an item about Muzak or conducting eight cars for a scrapyard car horn concerto should give you a flavor of such stunts. As a grand finale to a new film titled *Music of the Spheres*, we approached the European Space Agency to ask whether astronaut Samantha Cristoforetti would listen to and review my Kepler-inspired space composition from the cupola of the International Space Station. After a rather surprising affirmative reply, the initial stages of the composing process were relatively easy.

Kepler helped write the piece for me—I used the notes and intervals he suggested for each planet and placed their entries into the piece relative to their distances from the sun. However, what took forever was finding a suitable sound for each planet. I eventually landed on the idea of taking single notes from classical recordings, then shaping and looping them to create pulsating, ethereal sounds. I added the sound of a pulsar recorded at Jodrell Bank Observatory in northwest England and the original bleep of Sputnik 1. The pressure was mounting though: this composition was to be the first specially written piece of music to be played in space, a not insubstantial five million people would hear it on TV, and there was the risk that Samantha Cristoforetti might hate it. After much angst and creative turmoil, I completed my fifteen-minute piece, titled (rather unoriginally) *The Music of the Spheres*. Though Cristoforetti said it was the perfect accompaniment to watch the earth go by, the program's studio presenter summed up the film with a withering "Well, that was nice" and moved on to an item about dishwashers.

Chandra and the Plasma

The Chandra X-ray Observatory telescope was launched from the space shuttle *Columbia* in 1999 on mission STS 93. As the earth's atmosphere absorbs X-rays, lifting Chandra above our skies allows it to detect rays that could be a hundred times fainter here on the ground. One of Chandra's early discoveries was from within our own Milky Way galaxy, an X-ray emission from Sagittarius A, a "nearby" supermassive black hole. More importantly for this book, in 2003, Chandra recorded "sound" waves, the result of violent activity in the vicinity of a supermassive black hole in the Perseus cluster. However, if no one can hear you scream in space, how can there be sound? Of course, Chandra didn't *hear* sound; it observed ripples in the hot gas around the black hole. The observations were made not via visible light but through X-rays, part of the electromagnetic spectrum.

German physicist Wilhelm Röntgen first noticed these unknown waves in 1895 and therefore named them X radiation, after the traditional symbol for an unknown quantity. The letter stuck, though in many countries, they are still known as Röntgen waves or Röntgen radiation – more of Röntgen later in the book. The frequencies of these waves are incredibly high, between 30 petahertz and 30 exahertz (30,000,000,000,000,000 Hz up to 30,000,000,000,000,000,000 Hz). On our Infinite Piano, one would have to walk just over nine meters from middle C—about fifty-five octaves up—to find them. These incredibly high frequencies are the waves that Chandra looks for. As hot gas spirals toward a black hole, it is heated up to about ten million degrees Celsius, emitting X-rays back out into space. And hundreds of millions of years after these rays left Perseus, they were intercepted by Chandra.

But that still doesn't explain the boom of a low B♭. Where

did this note come from? Recent research suggests that, contrary to popular belief, not quite everything is sucked into a black hole. Let's imagine a galactic bathtub being drained of water. Everything is traveling toward the plug hole—water, soap suds, unwanted hair, old Band-Aids, and so on. In the chaos of super-fast spinning water and bath detritus, a vertical jet of water is sprayed toward the ceiling. In black hole terms, this jet would not be a light spray of slightly soapy, hairy water so much as a searing ejection of relativistic plasma traveling at almost the speed of light. Picture that vertical jet pushing its way through the hot steam of the bathroom. It is the *collision* of the jet of plasma with the surrounding gas clouds that creates booming shock waves and ripples, the source of the low B♭ "note" observed by Chandra.

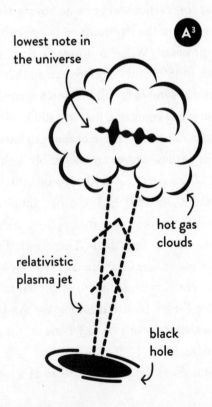

lowest note in the universe

A³

hot gas clouds

relativistic plasma jet

black hole

This isn't just any old low note though. This is, at present, the deepest pitch in the universe. Observing these ripples in the hot gas clouds, astronomers have calculated that the period of one of these shock waves is about one cycle every 18.5 million years. To understand a cycle, picture a drum skin being struck from above. (For ease, think of this in slow motion—slow motion of the drum skin that is. Do not try thinking more slowly.) The cycle starts the moment the stick makes contact with the skin. The skin is depressed, springs back up past its original position to a high point, then returns downward. When it passes back through its original position, it has completed one cycle. The skin will naturally continue to bounce down and up, but it is the length of time it takes for this initial down-up-down motion that defines its frequency. Using the standard unit of one cycle per second (1 Hz nowadays), the frequency of NASA's lowest note in the universe is 0.000000000000002 Hz. On our Infinite Piano, how many light-years to the left of middle C do you think you would have travel to play this mother of a sub-bass note? In fact, it's about the same distance from middle C as the X-rays are, only in the opposite direction—a slightly underwhelming 9.4 meters. And until another telescope spots a shock wave with a period longer than 18.5 million years, B♭ will remain the lowest note in the universe.

Space Music

Although such extreme notes have never been available to composers, many have attempted to portray the vastness of space using instruments that have the capacity to produce frequencies close to the lower limits of human hearing. Perhaps the most famous piece of space music is the opening section of Richard Strauss's *Also sprach Zarathustra*, which became the theme to Stanley Kubrick's film *2001: A Space Odyssey*. It begins with an ominous low C on a collection of rumbling instruments, including contrabassoon, organ pedal, and double basses. Some of the double basses play a C that is actually out of the range of a standard orchestral double bass, requiring a special extension to get down to this frequency of 32 Hz. (They're available to buy on the internet for a modest four-figure sum—thanks, Strauss.) The organ pedal is written at the same pitch, but with the right type of pipe, its C could potentially be an infrasonic rumbling 16 Hz, a feeling more than a sound. (Infrasound is defined as audible frequencies beneath 20 Hz, the lower limit of human hearing.) Although 16 Hz is not quite in black hole territory, this piece of music was undoubtedly an inspired choice on Kubrick's part. What other great works can boast infrasonic rumbles, anticipatory trumpet fanfares, arresting orchestral crescendi and the most brutal gear changes between major and minor tonalities, all within a matter of a few bars of music?

Though Strauss may never have dreamed that this work would become so closely connected with the cosmos, a contemporary of his, Gustav Holst, had the universe very much in mind when he composed the most influential piece of descriptive space music, *The Planets*. What is notably striking about Holst's suite, written between 1916 and 1919, is that it remains the go-to musical language used to portray all things astronomical. *The Planets* is impressive for many reasons, but particularly because it defined a cosmic musical language long before a single photo was taken in or from space.

One might imagine, having been presented with the astonishing images captured by Apollo, Mir, the space shuttle, and Hubble, that modern musicians would have invented a new, more informed musical style. But James Horner's homage to (or even occasional borrowing from) *The Planets* in his score for the 1995 film *Apollo 13* proves how indebted composers remain to Holst's creative imagination. Compare the similarities between Horner's "Docking" and Holst's "Neptune, the Mystic," both centered around two chords a third interval apart, both relying on diminished seventh chords, and written for similar timbral forces, including harps, flutes, celesta, glockenspiels, and wordless female voices. Echoes of Holst continue to appear in the works of even more contemporary composers such as Hans Zimmer. The connections between "Murph" from the film *Interstellar* and Holst's "Saturn, the Bringer of Old Age" are relatively easy to spot. Both contain instrumentally ambiguous, repeated burbling midrange accompaniment patterns, overlaid with soaring, slow-moving violin melodies. It seems that with only a pencil, paper, and a vivid imagination, Holst's musical expression of weightless serenity is yet to be bettered.

But neither Richard Strauss nor Gustav Holst were the first to portray space through music. Many argue that Austrian composer Joseph Haydn is the true founder of the genre, its big bang emanating from *The Creation* of 1798. The opening of the work, titled "The Representation of Chaos," is an imaginative tour de force, the sound of infinite nothingness. It depicts the beginning of the universe, the emergence of primordial light, and, in doing so, pushes many of the fashionable eighteenth-century conventions of harmony to the limits of acceptability. After the opening orchestral unison, Haydn's instruments move gingerly by step, often chromatically (using notes that do not fit the key of the piece), lingering unexpectedly into chords in which they have no place, creating ambiguous harmonies and unpredictable progressions, all of which go to portray the primordial soup that Genesis's opening verses describe. To this day, chromaticism and ambiguous harmonic relationships still dominate much of our idea of descriptive space music.

Ironically, though, one place where all these pieces deserve to be heard is *in* space, so it seems a missed opportunity that the discs of music carried on spacecraft intended for alien life-forms to find do not contain Haydn, Strauss, or Holst on their playlists. Back in 1977, *Voyager 1* and *Voyager 2* blasted off from Cape Canaveral bound for the space beyond our solar system. Just in case they bumped into extraterrestrials on their interstellar journeys, gold-plated gramophone discs were stored on board, containing human and natural sounds from Earth. The great cosmologist Carl Sagan headed up the team chosen to curate the disc. After the initial spoken greetings and the twelve-minute "The Sounds of the Earth" track, the first piece of music that our alien friends will be treated to is the opening movement of J. S. Bach's Brandenburg Concerto no. 2. Sagan was an intelligent, cultured man whose first choice to represent the pinnacle of human musical achievement I would endorse unequivocally. Also included on the golden record are sounds such as volcanoes and earthquakes; animals such as dogs, birds, and crickets; spoken greetings in fifty-five different languages; music from Senegalese percussion to Peruvian panpipes, Chuck Berry to Ludwig van Beethoven; and photographs as wide-ranging in subject matter as a page of Sir Isaac Newton's book *Philosophiae naturalis principia mathematica* and a woman eating what looks like a Ferrero Rocher chocolate in a supermarket. But of more interest to us are the recordings that one of the *Voyager* crafts itself captured and beamed back to Earth.

Voyager 1 is fitted with an instrument that detects changes in the solar wind. The instrument faces back toward the sun and measures the density of plasma—waves of ionized gas that come rushing past (the plasma is traveling a lot faster than the space-craft). The frequency of the waves tells scientists about the density of the gas around *Voyager*. During the first half of 2012, with the spacecraft at the edge of the heliosphere—the sun's bubble of influence—waves of plasma were measured around 300 Hz, the E♭ above middle C, which is the first note blown by the soloist at the start of Haydn's Trumpet Concerto in E♭, or Frank Sinatra's final uplifting note at the end of "New York, New York." Not long after, the frequency surprisingly jumped to between 2,000 Hz and 3,000 Hz. This change in frequency of the plasma indicated that the first ever human object had left our solar system and was now in interstellar space. If you were to place your right thumb on a piano C_7, your little finger would fall on the G above, around 3,000 Hz, and within your five fingers are the plasma notes of deep space.

Electrical Disturbances Apparently of Extraterrestrial Origin

Until the twentieth century, we were not even aware that we are surrounded by celestial bodies and events that are drenching us in all kinds of waves across the electromagnetic spectrum. Our ability to *see* space outside the visible spectrum was accidentally discovered by a man who wasn't an astronomer, and even after his discovery, he was still ignored for many years. Now known as the father of radio astronomy, Karl Jansky was a physicist and radio engineer. While working for Bell Telephone Laboratories in the late 1920s and early 1930s, he was asked to investigate the source of substantial amounts of static hiss and crackle that were interfering with long-distance transatlantic phone calls. As these used radio waves, such calls were prone to large amounts of extraneous noise, often rendering conversations impossible. Bell Labs was quick to realize that customer satisfaction would nose-dive if a seventy-five-dollar New York to London call was dominated by three minutes of a sound similar to that of frying bacon. They asked Karl Jansky to find out what was making the static hiss.

Jansky set about building a large antenna in New Jersey. But this was no vertical antenna or dish like the earthbound radio telescopes we see today. It looked like the Wright brothers' first airplane, a series of flimsy boxes on Ford Model T wheels, which allowed the structure to rotate—the antenna became known as "Jansky's merry-go-round." Jansky set the machine to receive a frequency of around 20.5 MHz (20.5 million Hz, around the

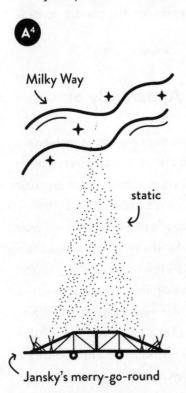

note D, three meters or sixteen octaves up from middle C on the Infinite Piano). What Jansky discovered was that most of the noise came from thunderstorms, both close and distant. But there was also another source that was more difficult to pinpoint.

The rise and fall of the static seemed to correspond with a cycle of day and night, but not quite—it was a few minutes short of twenty-four hours. It was only at this point that Jansky realized that the static corresponded to the night sky. He slowly narrowed the location of the static to a spot in the

Milky Way galaxy, in the constellation Sagittarius. In 1933, he published a paper on the subject titled "Electrical Disturbances Apparently of Extraterrestrial Origin." Spine-tingling stuff.

We now know that these radio emissions from near the center of our galaxy are the result of electrons accelerated in the strong magnetic field found there. But as many astronomers at the time thought Jansky was a bit bonkers and Bell Labs were not going to be able to stop a source of static emanating from far off in the Milky Way, very few scientists seemed interested. (The discovery was nevertheless well publicized, even making it into the New York Times.)

Only a few followed up on Jansky's work: another amateur, Grote Reber, who single-handedly built a radio telescope in his backyard in 1937; and Professor John Kraus, who initiated a radio observatory at Ohio State University after the Second World War. Without these pioneers, who knows whether we would ever have perceived the cosmic rumbling of the Perseus galaxy's lowest known note in the universe.

The Schumann Resonance

We do not have to look into deep space to find very low infrasonic waves. Here on Earth (or just above, to be precise), there is a low frequency that resonates right around our planet, excited into action by lightning. NASA suggests that at any given moment, around two thousand thunderstorms are active around the planet, producing about fifty flashes of lightning per second. The frequency these flashes stimulate has gained almost mythical status—it is claimed to have a positive effect on everything from our cognitive abilities to our blood pressure, from our heart rates to our gaits.

In 1952, Winfried Otto Schumann predicted the existence of a global resonance through mathematical modeling. Other scientists spent many years laying the foundation for his work through their research into the ionosphere, a layer of the earth's upper atmosphere between sixty and one thousand kilometers in altitude. Among many other things, they discovered that the ionosphere acts like a giant mirrored ceiling for electromagnetic waves, reflecting them back toward the earth. Schumann predicted that the ground and ionosphere create a vast chamber that extends all the way around the globe. At 7.83 Hz, the chamber becomes a particularly resonant cavity for electromagnetic waves. For this "tube" and the particles inside it to vibrate, there needs to be an initiation of some kind, just as when a drumstick strikes a drumhead. The stick in this instance is a five-kilometer plasma channel from which comes a one-billion-volt lightning strike that lasts for about 0.2 seconds and has an average temperature of around 30,000 degrees Celsius.

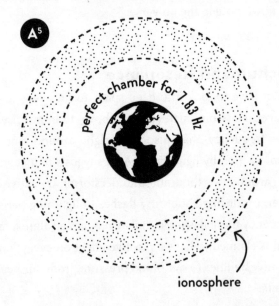

ionosphere

It is estimated that there are around 1.4 billion lightning flashes, in all their different forms, per year. But don't get confused at this point. The mystical 7.83 Hz frequency is not resonating air; it has nothing to do with a thunderclap, and it's not a sound. The lightning excites the electric and magnetic fields in our atmosphere, disturbing particles around our planet. Particle oscillation occurs in a large part of the frequency range, but our atmosphere between the ground and the ionosphere—our resonant chamber—amplifies 7.83 Hz better than any other frequency. There are related frequencies (overtones) that also resonate particularly well: 14.1 Hz and 20.3 Hz. But where would these frequencies sit on our Infinite Piano? The magic 7.83 Hz, just like the lowest note in the universe, is around a B pitch, but although it's well beneath the range of human hearing, it is still a whopping fifty-two octaves *above* the Perseus galaxy note. A question that astrophysicists have pondered, post-Schumann, is whether resonant frequencies are unique to *our* atmosphere. In theory, all atmospheres should have measurable frequencies resonating through them, and discovering and monitoring these could give us further insight into distant worlds.

With this in mind, the unmanned *InSight* spacecraft landed on Mars on November 26, 2018, in the exotically named Elysium Planitia. *InSight*'s mission has been to study the deep interior of Mars, to help scientists understand how the rocky planets of the inner solar system have been formed. Among the host of high-tech equipment onboard *InSight* is a Spanish-made weather center called TWINS (Temperature and Winds for *InSight*) and a SEIS (Seismic Experiment for Interior Structure)—they do like a good acronym at NASA. TWINS allows us to check daily reports of the weather in the Elysium Planitia region on Mars (yes, you really can), though it's an odds-on bet that there won't be a forecast for light showers in the afternoon. One of the early unplanned results of the mission was the discovery of the sound of Martian wind. There is obviously no breathable air on Mars, but this does not mean it has no atmosphere or other gases swirling around the planet. Even before SEIS was deployed into the Martian soil to begin its job of detecting "Marsquakes," it picked up the sound of a southeasterly breeze of about 16 miles per hour.

Lead scientist on the mission Bruce Banerdt explained why a seismometer is so important to the *InSight* mission: "An earthquake is almost like a little flashbulb. It illuminates the inside of the planet with seismic waves, and the seismometer is like a camera that picks up those waves and helps put together a picture. Pixel by pixel, we get to put together a 3D picture of the inside of a planet." After SEIS was deployed, it didn't take long for results to be heard and seen back at mission control. On April 6, 2019, *InSight* recorded its first Marsquake, a rumble centered around 8–9 Hz, beneath human audible range, but a strikingly eerie sound when sped up by a factor of sixty—not unlike someone blowing over the top of a half-full wine bottle in Westminster Abbey.

InSight is not the first mission to measure seismic activity on another world. Contrary to popular belief, our Apollo trips to the moon were not entirely peaceful and reverential. In fact, we did some damage. After placing a small seismometer into the dust around the Apollo 11 lunar module (LM), Neil Armstrong and Buzz Aldrin let off small explosions to measure the effect on their instrument. On Apollo 12, following the LM's safe return of the astronauts to the command module, it was unhooked and dropped back on to the moon's surface, creating a shock wave equivalent to one ton of TNT. A few months later, Apollo 13 catapulted the third stage of its S-IVB rocket at the moon. The rocket smashed into the lunar surface with a force of eleven tons of TNT, as measured by Apollo 12's seismometer, which had to be dialed down because of the force of the blast. The resulting shock wave lasted for three hours and twenty minutes. Apollo's seismograph network recorded over thirteen thousand seismic events and delivered some of the most important scientific results of the missions. It enabled scientists to better understand what the interior of the moon is made of—from Apollo's instruments, they deduced it has a solid inner core, though not made of green cheese, which settles the old argument.

The Great Football Match in the Sky

Focused more on *our* planet rather than the moon or Mars, the International Space Station has been taking photos and videos of Earth for almost two decades. Some of its most iconic images have undoubtedly been of the aurora borealis and aurora australis, which are among nature's most dazzling visual displays. But for centuries, the *sound* of the aurorae has swirled through the oral histories of peoples in both the Northern and Southern Hemispheres.

In 1877, the *Brisbane Courier* published a newspaper article titled "The Drought," which talked of recent aurora australis events: "There was a splendid Aurora in 1847, grand in its effects at Hobart Town; and an interim one on September 4, 1851, at the same place; where the vividly shooting streamers of violet, red and other colors, where [sic] somewhat marred by the bright moonlight. The aboriginies [sic] of Tasmania compared the crackling noise of the corruscations to the snapping of their fingers."

And in the earth's furthest points north, among the numerous accounts of aurorae within indigenous communities, reports of similar sounds were noted. In 1916, explorer Ernest W. Hawkes published his book *The Labrador Eskimo*, in which he recounted many of the stories and traditions of the indigenous population of northern Canada, including their explanations of the aurora borealis: "The spirits who live there light torches to guide the feet of the new arrivals. This is the light of the aurora. They can be seen there feasting and playing football with a walrus skull. The whistling crackling noise which sometimes accompanies the aurora is the voices of these spirits trying to communicate with the people of the earth. They should always be answered in a whispering voice."

In 1932, legendary Danish-Greenlandic explorer Knud Rasmussen also wrote about the great football match in the sky. He said that the Inuit peoples of Greenland believe "it is the ball game of the departed souls that appears as the Aurora Borealis, and is heard as a whistling, rustling, crackling sound. The noise is made by the souls as they run across the frost-hard snow of the heavens."

The really strange thing is that the thin air of the ionosphere where the lights are generated, up between 100 and 320 kilometers above the earth, is incapable of carrying sound waves. Scientists have studied the aurorae for decades, but the mystery of their sound continues. Some claimed for a while that it was caused by pine needles acting as tiny lightning conductors, but more recent research by Unto Laine from Aalto University in Finland found that the sound heard by Inuit and Aborigine alike is generated in the air, just over sixty-one meters above the ground. In 2016, his team suggested that an inversion layer in the atmosphere might be the cause of the crackling and whistling. This is where a layer of cold air, containing negative charge, sits beneath a warmer layer of air with a positive charge. During geomagnetic storms, the charge is released, creating the snap, crackle, and pop that the peoples of the north and south have reported for eons. We are still awaiting conclusive scientific proof.

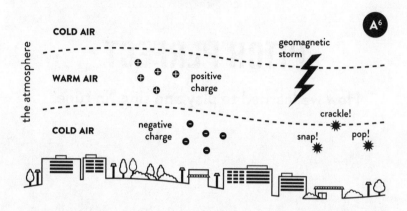

If Gustav Holst was composing *The Planets* today, around a century after its first performance, from where might he draw inspiration? Would he perhaps score the snaps and crackles of the sun's electromagnetic particles as they dance across the sky? Feasibly, he could add a few rumbles of Marsquakes (though they'd be inaudible to our ears) or enhance our well-being with the Schumann resonance frequency of 7.83 Hz and its overtones (again, most of which we cannot hear). He could even turn to Johannes Kepler's *Harmonices Mundi* ratios, connecting the planets and music—though he would soon receive a stern email from me, reminding him that this was my idea.

♭

PITCH PERFECT

How we learned to play and sing "in tune"

Reporting on the historic events at the Treaty of Versailles in 1919, the *Sydney Morning Herald* declared that "the occasion... was de'void [sic] of military character. Only seven regiments of Cavalry and four Infantry participated, without bands." The lack of musicians might not seem of particular significance to the occasion that sealed the end of the First World War. But the inclusion of a military band might well have opened a large can of diplomatic and musical worms.

Imagine a rather overenthusiastic bandmaster, swept up in the peace fever, excitedly suggesting a joint French, German, and British massed band to provide a moment of musical entente cordiale outside the Hall of Mirrors. The resulting performance would have been far from a stirring, euphonious anthem for peace. More likely, the combined national bands would have produced a catastrophic cacophony. You see, in 1919, there was no international standardized pitch to which all the individual bands could have tuned. For centuries, different countries (even different towns) each had their own ideas of which frequencies corresponded to which musical pitches. Following protests from singers and instrumentalists, in 1859, the French government established a standardized pitch for A above middle C at 435 Hz. (At this time, Heinrich Hertz was only two years old and had not been given the honor of a unit of measurement, so hertz was known as cycles per second—the unit of measurement, that is, not the two-year-old boy.) But A at 435 Hz only applied in France, and even after Queen Victoria officially announced in 1885 that her private band would adopt the French standard frequency, the British military continued to tune to "high pitch," where an A could be above 452 Hz. This lack of pitch parity meant that any collaboration between bands of different nations would have resulted in an out-of-tune mess. And then the recriminations would have started...

"You're sharp!" shouts a French cornet player.

"No, we're not. *Du bist* flat!"

"If you had adopted our French pitch agreed at the Vienna conference of 1885, then this would have all worked!"

"We're certainly not going to be dictated to by French musicians with your *flat* pitch, old boy!"

This leads to scuffles heard inside the Hall of Mirrors, the Germans refuse to sign the treaty, and the First World War continues for many more years.

Fortunately for the world in 1919, the imaginary bandmaster never foresaw the massed band opportunity, and the Treaty of Versailles was signed and the war was officially over. But the war over standardized musical pitch was most definitely not, and in May 1939—only a few months before the outbreak of the Second World War—a conference was held in London to discuss the adoption of A at 440 Hz as an international standard measurement. Influential voices at the event included the International

Broadcasting Union, the International Consultative Committee on Telephony, the BBC, and the British Standards Institution. The needs of international broadcasters and the rapid progress in technology were now pushing the musical agenda. In the early 1950s, as the International Organization for Standardization (ISO) prepared to validate the 1939 decision for an A to be set at 440 Hz, French composer Robert Dussaut campaigned against the pitch, claiming it had German origins. But the Americans had been using A 440 Hz since the 1930s, and as European instrument makers rushed to export trumpets, saxophones, and the like to the United States to make hay while the jazz sounded, the 1955 ISO conference was persuaded by 440 Hz. However, Dussaut had planted a seed that has grown into a conspiracy theory still espoused today, that Nazi propaganda minister Joseph Goebbels insisted on the "German" A of 440 Hz in 1939 in order to brainwash the music lovers of the world with an unnatural pitch standard.

An Apple of Discord

Our internationally standardized allocation of named pitches to set frequencies means we have jettisoned the vast majority of other frequencies into musical oblivion (in Western music at least). With this in mind, I've held on to an unproved theory: that one global tech giant has exploited our unfamiliarity with such long-forgotten frequencies in order to create an instantly recognizable aural logo for its brand. If the world standardized fabric colors, would you not dye your T-shirts shades in between these set colors, giving your brand unique recognition?

The company is Apple, and the genius aural logo is the start-up chime of an Apple Mac computer. It has a strikingly unique sound that most of us instantly recognize, though few can explain why; it is, after all, just one simple synthesizer chord. One clue to its singularity is that it's out of tune. It's not a C chord, an A chord, or a D chord. Its most recent variant is neither an F major chord (F_4 is 349.2 Hz) nor an F♯ major chord (369.99 Hz); the F of the Apple chime is somewhere in between the two. If all the music we listen to is bound by a series of set frequencies, then surely the way to make an aural impact is to place one's sounds outside these numbers? And Apple's start-up chimes are incredibly successful at expressing the company's identity in a single musical chord—thinking outside the box is definitely part of their core identity.

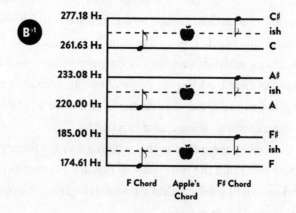

I was therefore excited to have the chance to interview the composer of the chimes, Jim Reekes. He is a remarkable and fascinating character, a composer, sound designer, software engineer, photographer, and *polterzeitgeist* (as he calls himself). Reekes told me that his idea for a start-up chime came from his need to be calmed whenever his Mac crashed. Before the famous chimes, Apple had installed cheap and nasty beeps that amplified his annoyance when his computer rebooted following a crash. These beeps were very much a product of the restrictions of the limited bit rate of computers and the inferior quality of their speakers. As the Macs improved, Apple employee Reekes saw an opportunity to compose a chime that utilized the full frequency range of their larger speakers. He wanted the sound to have a Zen-like "cleansing" effect, so he chose a C major chord, the simplest of harmonies, played on his Korg Wavestation synthesizer. But the chime was far more complex than it sounded. It consisted of notes that were organized ("voiced" is the musical term) in the same order as the harmonic series, ending with an E at the top.

"I added a third at the top and it just rung nicely with all the other partials. It made it a bright tone. Psychologically, we are hearing that and it doesn't sound resolved. It felt like you were lifting up, which I was intentionally trying to do." Reekes also spent a considerable amount of time experimenting with the sound of the chord, adding all kinds of effects, including chorus (making the sound richer and "fatter") and reverb (placing the sound in an ambient "space," as if in a hall).

I then asked the question that would prove my theory about the genius of Apple: the novel idea of placing the start-up chime in between established musical frequencies, giving it singularity in a noisy aural ident market.

"So how did you detune the whole chord by a quarter tone in order to make the chime so distinctive?" I asked. "Did Apple say, 'It's a great marketing thing; we're gonna stick it in the cracks of the piano where nobody else occupies that frequency space'?"

Reekes shook his head and grinned.

"You're overthinking it," he responded. "As far as I've been able to piece this together, it's a f***up. I wasn't there, so they probably got confused... When they played it back, they weren't musicians, they didn't know, they're f***ing engineers! They're like, 'Hey, sounds good enough to me.'"

Was it possible that this was a design fault on the part of Apple? Could a company that prides itself on such attention to detail have created a unique sound by accident? Reekes offered a number of different thoughts about the change in pitch, but a deliberate act on the part of Apple in order to give itself a unique musical signature was not one of them.

It was time to don my frequency sleuthing hat. In my studio work, I have occasionally encountered a problem where clients' audio waves (called stems) play out of tune (and a little faster) when loaded into my computer. The problem is caused by a sample rate discrepancy. To understand this, imagine a film made up of twenty-four frames per second playing through editing software set to play at twenty-five frames per second—the computer would have to stretch the film file, spacing out the frames slightly, meaning they would be slowed down in order to fit. This is the problem I sometimes encounter when clients send stems recorded at 44.1 kHz; played on my machine, which defaults to 48 kHz, they sound slightly lower pitched and play fractionally slower. The solution is to convert the clients' stems to the same sample rate as the audio software.

In February 1993, Macintosh released the Centris computer with its distinctive G major chime (which corresponds to the standardized G of 195.9 Hz). I recorded this start-up into my software set at 48 kHz. I then changed the project to 44.1 kHz—predictably, the chime sounded lower. But amazingly, its pitch dropped to exactly the same frequencies as the most up-to-date chime (which was introduced on the Power Mac 9500 in 1995). It appears Reekes was on to something—the random detuned chord found on all modern Macs can be achieved through a straightforward sample rate change. Does such an action seem deliberate or merely the slip of a mouse?

We went on to compare notes on our lives as composers. I commented on some recent rather banal generic music I had written for a TV documentary.

"Hey, at least you got paid!" Reekes said.

"Barely," I replied.

"More than I did," said Reekes. "Wanna see a grown man cry?" He never received a cent for his start-up chime.

Digging a Pitch Fork

Long before we officially elevated certain frequencies to musical stardom in 1955, an invention way back in 1711 was the first step toward musical standardization in a world of frequency chaos. A trumpeter, lute player, and instrument maker at the British royal court, John Shore, reputedly invented the tuning fork. Shore was one of George Frideric Handel's top trumpeters, until he "split his lip" and was "ever after unable to perform" (presumably on the trumpet, as it would be a poor excuse for never playing the lute again). Up until this point (and, as we now know, for many years afterward), musical pitch and its relationship to frequency were a complete lottery. A choir performing in 1710 at a venue without a fixed pitch instrument—such as an organ—might simply rely on a hummed note from their conductor. The inherent problem with this rather random system was that if the hum was too low, the bass singers might struggle to sing the lowest notes in their piece. Conversely, if the opening hum was too high, any higher passages in the work could lead to screeching sopranos, tensing tenors, and a rash of throat injuries. John Shore, ever the ex-trumpet-playing joker, told audiences that he always carried his "pitch fork" with him to aid his lute tuning, but in the process, he gave the world a fixed pitch that was portable, a set frequency that everyone could tune their instruments and voices to.

To celebrate Shore's revolutionary musical invention, I made a film about the tuning fork for BBC's *The One Show*. Handel's personal tuning fork was carefully transported to a Sheffield church—the location for our film—for us to shoot a short sequence featuring this historic musical relic. Its minder was a staff member the Foundling Museum in London, the home for this musical treasure. With obligatory white gloves, I picked up the fork from its box, keenly watched by the TV crew and the smiling Foundling lady. She had a polite, supportive smile, but her eyes said, "If you drop this fork, I'll stick it so far up your nose, your eyeballs will vibrate at 440 Hz." The director, perhaps unaware of quite how much of a world musical treasure this artifact is, asked the Foundling lady if I could strike the fork and sound it on one of the pews, so that we could hear the same note that the great Handel had tuned to all those centuries ago. Her eyes said, "Oh my God, no," but her mouth said, "Please be careful." I gingerly tapped the fork on my knee, expecting it to explode into a thousand tiny pieces that we'd have to sweep up off the church floor. Thankfully, it didn't explode, but when I rested its handle on the pew for it to be heard, the volume was too quiet for the microphone—I hadn't hit it hard enough.

"Would you mind if we strike it again?" asked the director.

"Okay, but I don't think we should do it many more times," replied the now very faint-sounding Foundling lady.

I knocked the fork harder this time while trying to maintain a cheesy BBC-magazine-show smile for the camera. It didn't explode, and I placed its end on the pew once more. This time, success. The fork sounded, and I was momentarily captivated,

sharing a unique firsthand musical experience with Handel himself. Unfortunately, it was cut short by the cameraman saying that
the camera had not been in focus. I could see the terror in the
Foundling lady's eyes, and she could see the beads of sweat on
my forehead.

After we'd got the shot on the third take, I had a few seconds
to reflect on a musical moment that was truly sublime, shortly
followed by the ridiculous—we had to set up for a recording of
Johann Pachelbel's Canon featuring a twenty-piece "forkestra." It
was time for the usual *One Show* musical stunt, a performance
combining the very silly with the mildly interesting and the awkwardly challenging. The forkestra piece was no exception. I had
arranged Pachelbel's masterpiece for twenty local musicians to
play on a range of differently pitched tuning forks. Only after
embarking on the arrangement did I realize how much of a logis-
tical as well as a musical challenge this was.

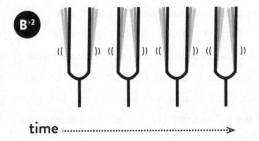

For a tuning fork to sound, it must be struck, setting the
two prongs of the fork (the tines) vibrating back and forth in
opposite directions. This was the first problem. All the players
had to have a "strike your fork" cue written into the musical
score prior to them placing the forks on a resonating surface.
Tuning forks are very quiet when vibrated in air, but their design
allows the vibrations to travel down the handle without one's

grip having much dampening effect. To get an audible sound from the fork, it is placed handle down on a resonant surface—the top of a wooden box, a pew, or even on one's teeth. For the forkestra's twenty or so performers, we opted to use three tables we found in the church.

The second problem is that it's very difficult to repeat notes in quick succession, as a performer must take the fork off the table, restrike it, and place it down again very rapidly—choice of music here was key. Two other challenges proved no less tricky. It's very difficult to read a musical score while hitting a small tuning fork in advance of its correct placement in the music. It's also quite a technical challenge, positioning microphones to successfully record a table that is very quietly resonating to the vibrations of forks without capturing the thump every time a performer placed their instrument down. I was very proud of the final film—to have combined an inspiring historical tale with twenty people banging forks on a table, all in a four-minute film watched by people eating their suppers on their laps while watching TV, is no mean feat. And no Handelian fork was harmed in the process.

The Rhodes to Recovery

My pride in "inventing" a forkestra was short-lived, as I soon found an intriguing story of someone who had beaten me to it. The man in question, a trainee architect, was forced to abandon his university studies during the Great Depression in 1929 to support his family. Curiously, he turned to piano teaching, a career not renowned for handsome remuneration. Aged nineteen, he took on his own piano teacher's pupils after she left the area, undoubtedly to find a better paid job. Harold Rhodes proved to be an inspiring music teacher, way ahead of his time, as his combination of classical methodology and jazz improvisation was a both popular and accessible way of mastering the piano. His teaching gained rapid success, and by 1930, he was managing a string of music schools across the United States, the Harold Rhodes School of Popular Piano.

When the Second World War arrived, Rhodes joined the U.S. Army Air Corps. Like every great teacher, he wanted to spread his passion, and he soon found himself teaching the piano to wounded soldiers. According to the *Los Angeles Times*, this apparently included instructing them in how to translate their girlfriends' phone numbers into melodies and accompany these tunes with simple chords. However, he quickly encountered a major stumbling block, quite literally, as his injured patients often struggled to get out of bed to the piano. The innovative and forward-thinking Rhodes developed a laptop piano, an instrument that a soldier could play while sitting in bed. He scoured his environment for piano-constructing materials and discovered that hydraulic aluminum pipes from the wings of B-17 bombers had interesting musical properties. Rhodes named the mini instrument the Xylette, and it quickly became a creative and therapeutic outlet for injured and bedridden GIs. The War Department saw its potential and, as well as inviting Rhodes to demonstrate his program at the Pentagon, manufactured thousands of these tinkly sounding laptop instruments. Even after receiving the Medal of Honor for his inspiring work with the military, Harold didn't sit on his laurels. He founded the Rhodes Piano Corporation, and the Pre-Piano was born in 1946. It was bigger than his Xylette but still only housed thirty-eight keys. Rhodes then fitted a microphone, amplifier, and speaker to the outfit.

On February 28, 1961, Rhodes was issued with U.S. Patent 2,972,922 for his "Electrical Musical Instrument in the Nature of a Piano," and the worlds of John Shore and Harold Rhodes converged. Although Rhodes's electric piano did not contain a set of John Shore's two-pronged (two-tined) tuning forks, it is the tines principle that connects the two inventors. In the patent that Rhodes filed in 1959, he explained that "another prior-art approach to the problem was to employ a large number of conventional tuning forks, but this was undesirable since conventional tuning forks have excessively long dwells." (Just like a bell, a tuning fork's ring takes time to die away, resulting in a muddy sound if a number are played consecutively.) The tines that Rhodes fitted to his Pre-Piano were, in essence, one half of a tuning fork, struck with a small hammer, and allowed to vibrate just like Shore's. The low acoustic volume of the single tine was overcome through amplification.

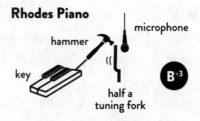

Rhodes twinkled on his merry way until he hooked up with iconic electric guitar manufacturer Leo Fender in 1959. But it wasn't until 1965, when Fender was bought out by CBS Instruments, that the Fender Rhodes piano took off as fast as anything at Cape Canaveral that year.

Throughout the late 1960s and the whole of the 1970s, popular music—whether it be rock, funk, jazz, anything—was dominated by the Fender Rhodes electric piano. The list of artists interviewed in Gerald McCauley and Benjamin Bove's book *Down the Rhodes: The Fender Rhodes Story* reads like a who's who of musical legends. And many of them highlight one of the greatest virtues of the instrument, in that the tone color it derives from the single tines makes it the perfect timbre to blend with other instruments. Conversely, its wide range of timbral colors makes it a fantastic solo instrument. Hit the keys hard and the notes are rich and spiky, dripping in overtones. Played softly, the tone is creamy and sensuous.

The Fender Rhodes was the perfect musical chameleon in its day, and Herbie Hancock's breakthrough "Chameleon" track on the album *Headhunters* is still one of the Fender Rhodes's most notable calling cards. A selection of others who have used the instrument to hit-making effect include the Beatles ("Get Back"), Miles Davis ("Bitches Brew"), Elton John ("Daniel"), Queen ("You're My Best Friend"), Stevie Wonder ("Isn't She Lovely" and a host of other tracks), Billy Joel ("Just the Way You Are"), the Doors ("Riders on the Storm"), Earth, Wind & Fire ("Let's Groove"), Bob James ("Theme from *Taxi*"), Bill Withers and Grover Washington Jr. ("Just the Two of Us")... I could keep going for a very long time.

Some of the instruments that were invented in the twentieth century soon dated—the Moog synthesizer, the wind synthesizer, the "keytar"—but the Fender Rhodes has continued to remain both retro cool and contemporary. Even though the early 1980s saw the rapid rise of the synthesizer, artists continued to use the Rhodes, from Madonna ("Holiday") through Radiohead ("Everything in Its Right Place") and Supergrass ("Mary") in the 1990s, through to artists of the 2010s such as Justin Timberlake ("Say Something"). Manufacturing of the Fender Rhodes ceased in the mid-1980s, and though every digital piano and computer music software package now contains a classic Rhodes sound, John Shore would undoubtedly agree with Harold that nothing beats the sound of a pitchfork.

All at C

Comparing the tuning forks of Handel and Beethoven, one is struck (pardon the pun) by the complete mess that prestandardized tuning really was. Though the two forks live less than a mile from each other in London, their tuning reveals a vast distance between them in terms of hertz. Beethoven's is kept at the British Library along with other treasures of his, such as scribbled kitchen accounts and his last laundry list. His fork resonates at 455.4 Hz, whereas Handel's C-pitched fork (the one I failed to shatter in Sheffield) resonates at 512 Hz, resulting in an A based on this sounding at approximately 422 Hz. The distance between Handel's and Beethoven's forks is 33.4 Hz, well over a semitone. Such was the tuning free-for-all that, around 1850, Broadwood piano manufacturers were issuing their tuners with three different forks for A: "low pitch" at 433 Hz, "medium pitch" at 445.9 Hz, and "philharmonic" pitch at 454 Hz. During the twentieth century, as the world got smaller, it is no wonder that calls for a standard tuning frequency became irrepressible. As with measures of weight, height, distance, volume, time, speed, power, luminance, and so on, our increasingly interdependent world needed a fixed frequency for musical tuning. By 1955, a world where A could be anything from Handel's 422 Hz to Beethoven's 455 Hz was just not practical.

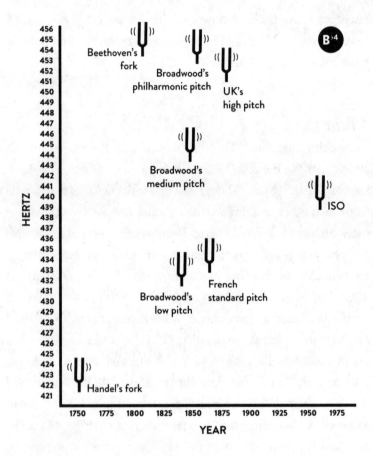

All of what I have described about the frequencies of tuning forks, though, assumes that manufacturers, scientists, and musicians can measure frequency, but until the mid-nineteenth century, they couldn't. Even then, the technology was in no way widely available. Broadwood's tuning forks all sounded a bit out of tune with one another, but they had no way of defining these small differences. So how did the accurate measurement of frequency come about? English polymath Robert Hooke conducted an experiment for the Royal Society in 1683 using his "sounding wheel." Hooke, who has been dubbed "England's Leonardo"

(da Vinci that is, not DiCaprio), had a startlingly diverse career, contributing to scientific understanding across a wide range of disciplines. It was a young Hooke who built the experimental apparatus that proved Boyle's law; Hooke's law of elasticity helped him develop the balance spring or hairspring that led to the first portable timepiece, the watch; he built a telescope and discovered a new star in the constellation Orion; he studied the crystal structure of snowflakes; he suggested that Jupiter rotated on its axis, and his sketches of Mars helped astronomers over a hundred years later calculate its rate of rotation; he was a professor of geometry at Gresham College; he studied the microscopic cavities in cork and coined the term *cell*; his work with fossils guided him toward a theory of evolution; he built a compound microscope with a screw-operated focusing mechanism; he discovered the phenomenon of diffraction (the bending of light rays); he discovered microscopic fungi; and to fill his spare hours, he had a sideline career as an architect.

In his 1683 sounding wheel experiment, Hooke was looking to prove the connection between pitch and frequency. He turned a toothed wheel that struck "an edge...a table, or a stone." Once the rotation of the wheel was accelerated to a point where the teeth made a continuous sound, the relationship between the frequency of teeth strikes and the pitch of the continuous note was discernible—the higher the frequency of strikes, the higher the pitch of the note. Though this wasn't necessarily new knowledge, it was a new and ingenious way of proving the principle. But unfortunately, Hooke stopped short of being able to numerically measure the frequencies of his spinning sounding wheel. It's possible he just got distracted designing new buildings in his role as surveyor to the City of London or drawing incredible illustrations of fleas' and flies' eyes for his 1665 book

Micrographia, or possibly he was too busy observing and drawing the rings of Saturn.

Hooke's Wheel

Robert Hooke's sounding wheels lay silent for over a hundred years until, in 1834, French scientist Félix Savart started turning giant brass wheels while holding a playing card or reed next to them. Savart extended Hooke's ideas by adding a mechanical tachometer (a rev counter not unlike the ones in our cars) to the axis of the wheel, and when calibrated, hey presto, specific tones could be associated with specific frequencies. He could listen to the pitch of the struck card, calculate from the tachometer the frequency of the teeth strikes, and give a specific numerical reading for that pitch. Robert Hooke must have been spinning in his grave (perhaps even on his own wheel) to think that his invention had now become known as "Savart's wheel." Hooke was renowned for being a curmudgeonly character who was perpetually concerned that he was not getting enough credit for his work. He remained in a long and bitter feud with Sir Isaac Newton right up to his death in 1703; they argued acrimoniously over the nature of light, and Hooke demanded unsuccessfully that he be credited as the author of the idea behind Newton's law of universal gravitation. Ironically, though, both Hooke and Savart were shortchanged by history, as Hooke's wheels became Savart's, and Savart never truly received the recognition for adding the key item, the tachometer.

Scientific Pitch

If Savart's wheel of 1834 was the first time in history that frequency was measurable, then we should reconsider the veracity of certain musical "facts." Journalists, authors, and reputable musical institutions often regurgitate stories about "perfect pitch" possessed by the likes of Handel, Mozart, Beethoven, and Chopin. This is the innate ability to produce or identify a musical pitch without any other reference point, a prized talent most musicians covet. But how could Mozart—who died in 1791, way before Savart's wheel—have sung a perfect A, B, or C when there was no way of measuring it? Was this self-promotion or approximation? Though Mozart couldn't prove he was singing a perfect A, no one could prove he wasn't. The only thing we can say with certainty is that the God-given, superhero ability of perfect pitch was invented in 1955, the year the ISO standardized A at 440 Hz.

It would appear that the argument is now settled, the war over pitch standardization has been won. Not so fast, International Organization for Standardization! Many believe that 440 Hz is a strained and unnatural number on which to base our musical tuning. Type 432 Hz into any online search engine, and you will find countless pages proclaiming the virtues of this frequency, known as "scientific pitch." Perhaps the least controversial claim is that it's a mathematically neat tuning standard—this is undeniable. Ditching A as the standard tuning note, the go-to frequency for advocates of the scientific pitch is C at 256 Hz. The C an octave lower would be a straight 2:1 ratio, 128 Hz. An octave lower again, 64 Hz, then 32 Hz, then 16, 8, 4, 2, and 1...neat. Using this method, with some tiny tweaks of temperament, our tuning A would be at 432 Hz. The composer Giuseppe Verdi was a supporter of this frequency for a time, and many singers claim it has a more "natural" feel.

Though this may be the case, not many singers I have met will stretch the claim to suggest that such a tuning system has healing powers. Many extreme devotees of scientific pitch insist that music should be flattened down to this frequency for the good of us all. Indeed, some claim that A at 440 Hz is "disharmonic" because it has no relationship to the universe. The "magic" 432 Hz's powers are allegedly related to the fundamental "beat" of planet Earth. This is claimed to be at 8 Hz, due to the very same Schumann resonance that we met in the previous chapter. It is also claimed that 8 Hz is the frequency of the double helix in DNA replication, and it is common scientific knowledge that some human brain waves are at 16 Hz. As 432, 16, and 8 have a simple mathematical connection, it's hard to see what other proof one could possibly need. If we can all simply get our brain waves to 16 Hz and focus on the Schumann resonance, our bodies and minds will resonate naturally with the rest of the universe. Some suggest that listening to music tuned to A at 432 Hz will fill us with a sense of peace and well-being, regardless of the kind of song chosen.

To test this theory, I listened to a slightly flattened version of that happy-go-lucky ditty "The Chicken Dance" with its A set at 432 Hz. I must admit that the effects were very subtle. It could possibly have been that the Schumann resonance isn't just one frequency, or that its fundamental is around 7.83 Hz and not 8 Hz, or that this 7.83 Hz is often affected by solar-induced perturbations in the ionosphere. Or perhaps I wasn't trying hard enough to focus my beta brain waves to precisely 16 Hz (one of my pet peeves is not being able to keep my brain waves at specific frequencies). But then again, it might just have been that the irritation value of "The Chicken Dance," even at 432 Hz, is more powerful than the "beat" of planet Earth and the frequency of the whole universe.

BRIDGES OVER TROUBLED WATERS

The vibrations of disasters

Almost ten years after the fall of the Twin Towers in Manhattan, a skyscraper in the South Korean capital of Seoul was evacuated after it started vibrating rather alarmingly. On July 5, 2011, the thirty-nine-story Techno Mart building shook for around ten minutes. Bizarrely, no seismic activity was reported in the area.

"I fled the building with everyone else while it was shaking up and down. It almost made me feel dizzy," Lim Joon-hee—who worked on the twentieth floor—told Yonhap News Agency. The *Korea Times* reported that "the fire station dispatched 23 rescue workers and seven fire trucks to help people exit the building and prevent others from entering." It also speculated that "a sports center located on the twelfth floor has been causing significant level of vibration throughout the structure as loud music is played there."

The authorities closed the skyscraper and investigated the cause. Was an earthquake to blame? Was it some kind of failed terrorist plot? Or was there a fundamental design flaw in the tower? Their conclusion was that the scary wobble of the building had been initiated on the suspected twelfth floor, but not exclusively by loud music. The culprits were identified as a group of twenty-three locals, devotees of an American guru called Billy Blanks. Though Blanks has a worldwide following, he is no terrorist. His harmless mantra and regime has, over many years, led millions to don Lycra, T-shirts, and headbands and work out to a sweaty blend of martial arts and boxing moves, accompanied by, yes, that loud, pumping dance music. Blanks developed his unique mash-up brand of boxing/martial arts/fitness in the late 1980s, and videotape sales reportedly grossed between $80 and $130 million. As the Tae Bo Twenty-Three—as I have named them—worked out to that 1990s classic hit "The Power" by Snap, it's ironic that

this small band of devotees was oblivious to the genuine power their moves contained. But not even a massed throng of Tae Bo fans should have been able to shake a skyscraper close to destruction.

The Tae Bo Twenty-Three had created a frightening chain reaction. Their combined dance moves caused the building to vibrate, initially by an imperceptible amount. However, their steps had inadvertently found the resonant frequency of the structure and were increasing its sway with every foot thump. They were unaware of the effect they were

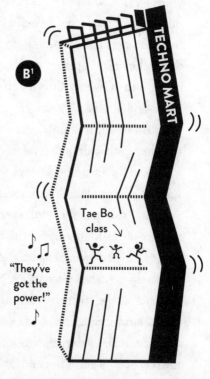

The building didn't actually sway this much!

having on the tower, as the size of the vibrations increased not on their own floor but on the floors above them.

This phenomenon is called mechanical resonance, a major consideration for all building designers, architects, and engineers. Unchecked, it can lead to resonance disaster, examples of which we'll explore later in the chapter. All objects vibrate, and the frequency and intensity of the vibrations are determined by a large range of factors, including the size, shape, and material of an object. The natural resonant frequencies ("sweet spots" of oscillation) can be intensified if an external force's vibrations are at or around the same pitch. The opera-singer-and-the-exploding-wineglass trick is a perfect example of how to create resonance disaster from an object's resonant frequency. The Techno Mart tower, like all buildings, is subjected to constant vibrations: occupants walking inside it, perhaps slamming doors; vehicles passing outside; the wind; the noise of planes overhead; a rumbling metro train below; or a group of twenty-three people working out. Structures all over the world deal with such vibrations with seemingly little or no problem at all. It is only when the frequency of such vibrations matches a structure's natural frequency that resonance is created.

A child's swing is always the go-to analogy at this point, so I won't disappoint. When a child clambers on to a swing and shouts "Push me," the adult exerts a force on the swing and child with said push. To amplify the resonance of the swing, the adult waits until the swing has traveled to the end of its return movement, then pushes once more. No sane adult shoves the child in the back as the swing is returning but instead pushes in sync with the natural frequency of the swing's movement, thus increasing the size of the swinging motion (amplitude) and the volume of the "Whee!" from the child. In essence, this is what the Tae Bo Twenty-Three were doing. Their movements were perfectly in sync with the movement of the Techno Mart tower. The Twenty-Three kept pushing the vibrating structure in time with its natural frequency, like the adult pushing the child in the back just as the swing moves forward again. And on they carried in their own sweaty way, probably until the commotion of hundreds of people rushing down the stairs led them to abandon their session.

Following the test results, engineers were brought in to stop any future vibrational shenanigans. Instead of simply banning Tae Bo on the twelfth floor, they went for a more scientific solution. The Techno Mart building was fitted with a hybrid mass damper, an instrument that is literally tuned to the natural frequency of the building. Techno Mart's damper is rather complicated in that it has two separate tunings (hence hybrid), one for vertical movement (Y-axis) and one for lateral movement (X-axis). Its Y-axis is tuned to 0.19 Hz and its X-axis to 2.7 Hz. The hybrid mass damper's movements work in opposition to the natural swing of the building, countering the skyscraper's motion. Through their analysis, the South Korean engineers discovered that the Techno Mart tower of Seoul gets vertically excited whenever it hears the opening G of Beethoven's Fifth Symphony (eleven octaves lower though) and laterally excited by the first F Kylie Minogue sings to "La, la, la" in "Can't Get You Out of My Head" (but seven octaves lower).

Tesla's Seismic Vibrator

Mechanical resonance has been known about for a long time, and one of the most brilliant and bonkers minds of the late nineteenth and early twentieth centuries put this knowledge to frightening use. The Serbian-American engineer/inventor/physicist/futurist Nikola Tesla had fingers in many scientific pies. His pioneering work included wireless remote control and power transfer, radio,

television, and fluorescent lights. Tesla also invented something
rather less well known but even more startling—an earthquake
machine. In 1894, he obtained a patent for a "reciprocating engine,"
a machine with pistons, in essence a steam-driven vibrator. He
hadn't set out to invent an earthquake machine; this alarming
gadget was a by-product of Tesla's experiments in the production of alternating current (AC). He developed a mini seven-inch
version of the reciprocating engine which "you could put in your
overcoat pocket," and in 1898, he attached it to a steel girder in
his laboratory in Manhattan. He later wrote, "I put the machine
up a few more notches. There was a loud cracking noise. I knew
I was approaching the vibration of the steel building. I pushed the
machine a little higher. Suddenly all the heavy machinery in the
place was flying around."

There are differing accounts of how the alleged localized
tremors were stopped, but what is indisputable is that panic
ensued, the local police were summoned, and Tesla smashed the
mini earthquake machine off the girder with a large mallet. In an
article in *The World Today* in 1912, he claimed his machine could
"drop Brooklyn Bridge into the East River in less than an hour." He
also boasted that it could level the Empire State Building within
ten minutes and that it could split the earth in around two weeks.
Tesla was probably more interested in causing a PR stir rather than
an earthquake, but could he really shake things up Richter-style?

Earthquakes release massive amounts of energy in very
small time frames—a magnitude 5.0 event releases energy
roughly equivalent to that of the Nagasaki atomic bomb blast. To
go from 5.0 to 6.0 requires a tenfold increase in the amplitude of
the waves. On the other hand, the vibrations caused by earthquakes are far more complex than simply the vertical motion of
Tesla's reciprocating engine.

As one might guess, seismic waves come in many different forms. There are ones that travel on the surface and ones that can penetrate through a body. Some go up and down, others move laterally, some are not supported by fluids while others can travel through oceans, and many are a mix of different types. The names of these waves range from the exotic, such as Love waves and Rayleigh waves, to the more mundane P waves and S waves. (In the simplest of terms, P waves cause the ground to move back and forth, in the direction of travel, whereas S waves travel in an up-and-down motion, as if shaking a rug.) Earthquakes produce a complex mix of these waves, so sorry, Tesla. You were never going to create an earthquake from your mere vertical seven inches.

Earthquakes also produce sound waves. Researchers have noted that human perception of seismic sound depends on the magnitude and distance from the epicenter, and, perhaps counter-intuitively, small seismic events are easier to hear than large ones. One of the problems that humans have with earthquake sound is that, in the main, it's too low for our ears. Large earthquakes tend to boom out in the infrasonic range, down where the Schumann resonance lives, generally between 1 and 10 Hz. On the Infinite Piano, the lower limit of these frequencies would be around 1.3 meters to the left of middle C, down about eight octaves. Though humans are incapable of hearing in this range, many animals can. And it is this ability that is now being considered as an explanation for the centuries-old tales of creatures who appear to sense earthquakes, volcanic eruptions, and tsunamis long before humans.

The Elephant That Never Forgot the Wave

During Christmas 2004, eight-year-old Amber Owen was enjoying a beach paradise holiday at the Sheraton Hotel in the resort of

Choeng Thale on the Thai island of Phuket. She had befriended Ning Nong, a baby elephant that was regularly brought to the hotel to give children rides. Ning Nong and Amber's bond was especially close, and Amber had ridden on Ning Nong's back every day of her holiday. On Boxing Day morning, she went for her usual ride down on the beach. Underneath Ning Nong's feet, the Indo-Australian tectonic plate was continuing its own twenty-million-year-old merry ride, very slowly grinding beneath the Eurasian plate at a rate of about five centimeters every year. But at 7:59 a.m. local time, terrifying forces of nature were unleashed, as the Eurasian plate—which was being bent downward like a plastic ruler off the end of a table—snapped back up between ten and twenty meters. The earthquake was more powerful than all the quakes of the previous five years put together, a terrifying 9.1 on the Richter magnitude scale. It released the energy of twenty-three Hiroshima-sized atomic bombs. That energy produced ocean waves that were soon traveling away from the epicenter at almost 1,000 kilometers per hour, faster than the speed of a jet airliner. And part of this tsunami headed for the Sheraton Hotel in Choeng Thale, five hundred kilometers away in Phuket, where Ning Nong was carrying Amber Owen on her morning ride.

From her vantage point, Amber could see that the tide had retreated far out to sea. The elephant's keeper was picking up the stranded fish on the sand, but the elephant kept pulling away from him, seemingly in an attempt to escape. As the first wave rushed in, Ning Nong ran off the beach and delivered Amber to a high wall where she could step off his back. Amber's distraught mother was soon reunited with her daughter, and both fled to a safer place before the larger second wave smashed into the shore. Ning Nong survived the further waves, and Amber's story became the inspiration for Michael Morpurgo's book and West End play *Running Wild*.

Ning Nong's pre-disaster "sixth sense" is far from the first on record. For centuries, people have claimed that animals' erratic behavior has proved to be the warning of tremors, tsunamis, and volcanic eruptions. The third-century Greco-Roman author Claudius Aelianus wrote of a mass exodus of animals from the Greek city of Helike, days before it was struck by a devastating earthquake and tsunami in 373 BCE. In 1975, the Chinese city of Haicheng became the first earthquake zone to be evacuated in advance of the tremors. Many lives were undoubtedly saved, and some suggest that this action was due in part to reports of animals' strange behavior over the weeks and even months leading up to the quake. Researchers noted that a substantial number of reports of snakes and frogs coming out of their dens into freezing temperatures during the hibernation season were difficult to ignore, as were the sightings of disoriented mice. (Doubt has been placed on a lot of these incidents, as the witnesses were interviewed *after* the quake. "Oh yes, now I come to think of it, my dog ate his dinner particularly quickly just before the earthquake" would not be counted as solid scientific evidence.)

Ning Nong has a slightly more compelling connection to earthquakes and tsunamis than disoriented mice though. Elephants are highly attuned to infrasound, vibrations beneath 20 Hz. Indeed, research is now proving that elephants hear through their feet as well as through the more obvious aural appendages on the sides of their heads. As well as hearing and feeling infrasound, they can produce it in their calls. Through both the air and ground, bass notes travel farther than higher frequencies, something you will have noted if you've heard a live rock gig from a distance— much more thumping bass drum than vocals. Studies show that low frequency and infrasonic elephant-generated seismic waves can travel up to thirty kilometers. Elephants communicate across large distances by producing infrasonic vocal rumbles. A leading researcher in elephant communication, Michael Garstang, suggests that in the right conditions (temperatures at around 5°C), elephants can produce calls as low as 10 Hz, an octave and a half off the end of a standard piano. As elephants have vocal cords up to eight times the size of human cords, this is hardly surprising. But to put this frequency into context, let's consider how big a tuba we would need to play such a note.

A typical double B♭ tuba—the lowest standard tuba in an orchestra—has around 5.5 meters of pipe, is about one meter tall, and plays a fundamental note at approximately 29 Hz. To get to 15 Hz, you'd have to manufacture a tuba with around eleven meters of piping, probably standing over two meters tall. To delve down to the elephants' 10 Hz, you'd have to purchase a small truck to transport your three-meter tuba with its eighteen meters of tubing. But what is even more extraordinary than a giant tuba is how elephants detect infrasonic sound. When "listening" to their contemporaries' calls, they squash the fatty part of their heel—called a digital cushion—down into the ground. Touch receptors in the same area transmit messages to the brain when detecting the vibrations of other elephants' calls. Caitlin O'Connell-Rodwell, another of the world's leading researchers into elephant communication, suggests that elephants triangulate signals using their feet and ears, thereby enabling them to determine distance, direction, and the content of incoming messages.

So is it possible that Ning Nong heard the infrasonic rumble of the tsunami long before it smashed into the shore? Could this explain why he tried to escape the beach? American and Sri Lankan scientists were studying the behavior patterns of two satellite-tracked collared Asian elephants on the day of the earthquake. Their data did not indicate any flight response from either elephant, both of whom were living close to the beach at Yala National Park in Sri Lanka. They found no animal "sixth sense" that the media reported in so many post-tsunami stories. What is indisputable, though, is that Ning Nong saved Amber's life. The reason for the miracle is open to much speculation. Did he hear the tsunami coming? Was he merely spooked by the sight of countless fish slapping around on the exposed sand? Or perhaps he simply didn't want to be on the beach at that particular moment.

Garstang and his colleague Michael Kelley reported that "substantial evidence exists that animals can detect a range of abiotic sounds [these are sounds not produced by living organisms, such as rushing water, wind, etc.]. Crustal movements in the earth's near-surface produce such sounds." Yet the route through which such sound could have arrived at Ning Nong's ears is most surprising. The authors cite one of their own previous studies, which suggests that crustal movements prior to an earthquake induce distortions in the earth's electric fields. The distortions are transmitted by the earth's magnetic field up from the ground all the way to the ionosphere at the speed of light. These electron fluctuations now in the ionosphere "are transferred to the earth's surface... At this point, they are not in the form of sound waves but trigger vibrations in metal, glass and other surfaces." Such electron fluctuations have an almost instantaneous transmission speed, and not only do they outpace all other potential warning signs of an earthquake, they can occur *before* the main tremors. With this in mind, it's just possible that Ning Nong might have been spooked by a low-frequency hum coming from the metal tables at the local beach café or an almost imperceptible rattle from the glass windows at the nearest hotel. Garstang and Kelley propose that "reliable response to the adherent animal behavior could still prove useful, particularly in poorly monitored remote areas."

The Supersonic Coo-dunit

A few years before the Boxing Day disaster, another infrasonic-laden miraculous tale concluded in Manchester, England. In 2002, a homing pigeon named Whitetail arrived back at his loft in the suburb of Hattersley. Whitetail was a racing pigeon, a champion that had won thirteen races and crossed the English Channel fifteen times. He was one of sixty thousand contestants in the Royal Pigeon Racing Association's centenary race. What was extraordinary about Whitetail's return in 2002 was that the race had begun in Nantes, France, at 6:30 a.m. on Sunday, June 29...1997. Instead of his anticipated return during that ill-fated Sunday afternoon, Whitetail had simply vanished, along with the vast majority of pigeons that had been released that morning. Even Queen Elizabeth II lost some of her birds in the race, an event that very quickly became an avian whodunit. Five years after the race had commenced, Whitetail nonchalantly sauntered back into his loft in Hattersley. His fancier, Tom Roden, said, "I was absolutely amazed. He must have a phenomenal memory to recognize his way home after all this time." Whitetail responded with a Gallic shrug.

But who or what had inflicted such a catastrophic pigeon-based calamity? How could so many experienced flyers, known for their legendary navigational skills, fast flight speeds of 50 miles per hour, and values of up to £100,000, simply disappear en masse? The "coo-dunit" might finally have been solved by an unlikely detective, research geophysicist Jonathan Hagstrum from the U.S. Geological Survey. His work in the field of animal navigation led him to metaphorically invite various suspects to sit on leather sofas in front of a roaring fire in an oak-paneled room, accompanied by an underscore of brooding music. After the obligatory opening of "You're probably wondering why I gathered you all here," he spun on his heels, stroked his impressive mustache, and pointed the finger of blame at...the Concorde, the supersonic airliner that whisked jet-setters around the world between 1976 and 2003. One of the shifty-looking suspects in the room challenged Hagstrum: "But there were no reports of bird strikes in northern France on Sunday, June 29, 1997, and sixty thousand pigeons colliding with the Concorde would have definitely made the news!" Hagstrum stroked his impressive mustache again, sighed, and embarked on his explanation.

He claims that pigeons can pinpoint their location and find their way home by listening to the infrasonic sound of the earth. He further suggests that these birds learn to memorize an acoustic map of their home location based on the infrasonic rumbles of the seabed and microscopic movements of the ground we stand on. It is indeed a scientific fact that there is constant infrasonic noise throughout the earth generated by ocean storms and movement of the sea. It is also a fact that we are not standing on rock-solid ground. The crust of the earth is constantly moving up and down, vibrated by microseisms—tiny oscillations transmitted from the deep ocean floor to the continental land masses, caused by oceanic storms and swells. On land that is close to a shoreline, the ground might move up and down by about four microns (four millionths of a meter), roughly the width of a strand of spider silk. Further from the ocean, the ground might only fluctuate by less than a micron (if at all), the length of a typical small-sized bacterium. But nevertheless, we live on a wobbly world. And that movement creates noise: infrasonic noise.

Seismic Matters

We've been monitoring seismic vibrations for a very long time. In 132 CE, the Chinese scientist and polymath Zhang Heng invented a seismoscope called an "earthquake vane." It could reputedly give the direction of the epicenter of a tremor by dropping a ball from one of eight dragons' mouths into a corresponding frog's mouth at the base of this ornate bronze instrument. More sophisticated seismometers were used at the Maragheh observatory in eastern Iran during the thirteenth century, and the birth of modern seismometers is claimed to have begun with Scottish geologist John Milne, who established an observatory on the Isle of Wight in 1895. From there, Milne managed a global network of seismometers and published his findings of seismic activity. This principle of global networks remains in place today, from the Global Seismographic Network, which draws together over 150 digital seismic stations, to the Comprehensive Nuclear-Test-Ban Treaty Organization, a system of over three hundred facilities monitoring nations' nuclear test explosions. Sixty of the organization's sites are specifically infrasonic listening locations, places that undoubtedly hear the mysterious sounds of our land as it moves up and down. But what frequencies are generated by ground that moves by millionths of a meter, and how on earth can pigeons hear this?

Professor Kiwamu Nishida of the University of Tokyo states that "since seismologists began observing seismic waves from earthquakes, the existence of an ambient seismic wave field with a dominant frequency of about 0.15 Hz was firmly established." (Its musical note is a slightly flat E♭. When transposed up eleven octaves, it is the start of the first riff of Jimi Hendrix's "Voodoo Child.") Other infrasonic frequencies resonate through our planet, but this particular frequency is the most prevalent. If we imagine that the ground we are standing on is a trampoline bouncing at 1 Hz, it would take one second for the mat to rise, fall, and return to its original position. At 0.15 Hz, just over a sixth of that motion would be covered in one second. In other words, the trampoline would be moving much slower and would take over six seconds to rise, fall, and return. It wouldn't be the most fun of trampoline bounces either, since the microseismic distance covered would be smaller than the length of a typical human spermatozoon's head. But though no child would be excited about the prospect of a trampoline bounce of six seconds across a distance of less than a human hair, the atmosphere is most definitely excited by it. Albeit by a tiny amount, the air above the trampoline surface is pressurized as it is squeezed by the almost imperceptible rise of the trampoline. This is how the infrasonic frequency is transferred from the ground to the air. As a trampoline's mat is usually flat, the pressure on the atmosphere is extended straight upward into the sky. But if one imagines that the trampoline is angled—like the side of a mountain—an infrasonic wave would then travel diagonally. Add in the factor that sound is bounced off certain layers of the atmosphere, and one begins to understand how infrasonic sound swirls around us, coming from all directions.

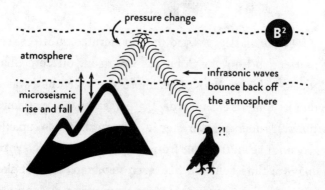

This is where Whitetail, the Concorde, and Jonathan Hagstrum rejoin us. Hagstrum believes that pigeons have the capacity to "hear" infrasonic frequencies such as 0.15 Hz, even though biologically, this seems impossible. When pigeons first take off, they circle around to get their bearings. Using Doppler shift (where sound traveling toward or away from a subject changes in pitch), Hagstrum believes they can build up an aural picture of these low frequencies and their very long wavelengths. And because infrasonic sound comes from all directions, Hagstrum suggests they quickly map their surroundings.

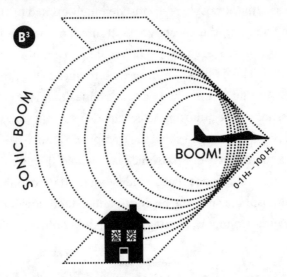

As Whitetail flew toward the northern coast of France on that morning in 1997, the Concorde was taking off from Charles de Gaulle Airport in Paris. This impressive speed bird was way ahead of its time (the Concorde, that is), an engineering miracle and the first supersonic airliner in the world. It was capable of transporting over 120 people from London or Paris to New York, arriving at a time before it had even departed. But not everyone loved the Concorde. Objections were raised as soon as the plane first took off, in particular because of its sonic boom. As a flying object travels, the sound it generates is projected outward like ripples in the center of a pond. In the direction of travel, the object will follow a ripple but never catch up with it, as long as it travels slower than the ripple itself. As the object is constantly emitting ripples (its noise), these ripples bunch up more and more in front of the object the faster it travels.

The problem arises when the object is traveling at the same speed as the ripple (the speed of sound). The ripples in front of the object are all bunched up into a high-pressure wave, which eventually goes "Bang!" When the object travels faster than the ripples (breaks the sound barrier), it produces a wake that is cone-shaped, a bit like the wake of a ship, but in three dimensions. The edge of this cone is a percussive pressure wave whose "bang" can be heard beneath the flying object. The cone travels along the ground, producing a continuous boom, as if the plane is dragging a skirt behind it. The problem for Whitetail and his sixty thousand rival pigeons was that a sonic boom's energy range is concentrated between 0.1 Hz and 100 Hz, the lower part of which covers the infrasonic spectrum that Hagstrum believes pigeons use for navigation. Due to complaints from those who lived under the flight paths of both the British and French Concordes, the jets were never allowed to accelerate past the speed of sound until out over water. In the case of the French Concorde, this meant flying subsonic until over the English Channel, when the pilot would put his foot down and break through Mach 1. If Hasgstrum is correct in believing that ocean storms vibrate the seabed, creating microseisms that travel through the ground up through mountains, vibrating the air in the atmosphere that helps pigeons create infrasonic maps that allow them to find their lofts in Hattersley, then a big bang of high-volume infrasound could have had a catastrophic effect on the navigation systems of the sixty thousand pigeons participating in the centenary race of the Royal Pigeon Racing Association. It would be equivalent to suddenly turning off all the navigation aids of all the drivers in London. Whitetail was one of the lucky ones, undoubtedly glad to be home, though now cooing with a French accent.

On his final flight, circling around Manchester to recognize his old surroundings, it's just possible that Whitetail flew over a bridge on the west side of the city whose own frequency story is just as intriguing and has had a direct impact on all military personnel since 1831.

A Bridge Too Far

Broughton Bridge was an early Victorian marvel. New suspension-style bridges were becoming all the rage, and the River Irwell was spanned in 1826 with a gleaming forty-four-meter chain suspension bridge, similar to its larger cousin over the Menai Straits between north Wales and Anglesey at a more impressive 417 meters. The design had been around since the fifteenth century, when Tibetan bridge builder Thangtong Gyalpo constructed a number of simple iron chain suspension bridges in Bhutan. But it wasn't until the nineteenth century that the real boom in their popularity occurred. Nothing seemed to dampen the public's appetite for this style of bridge, even though there were many failures, including Dryburgh Abbey Bridge in the Scottish borders, which was constructed in 1817. It was an impressive seventy-nine meters long, but less impressively, it collapsed the following year (as did its replacement in 1838). However, the gleaming new Broughton Bridge, which connected the two villages of Broughton and Pendleton—both of which are now within

Greater Manchester—was certainly not going to succumb to such an ignominious fate. Locals were incredibly proud of this engineering marvel, and for five years, the bridge did its job, carrying pedestrians and vehicles back and forth, inspiring a newfound bond to blossom between the two villages.

Alas, though, pride in the bridge (along with the bridge itself) collapsed on April 12, 1831. Soldiers from the Sixtieth Rifle Corps were returning to barracks in Salford after military exercises on a local moor. They approached from the east, marching four abreast, undoubtedly happy to have completed about half of their homeward journey. As they crossed Broughton Bridge, it began to vibrate, which amused many of them. Merrily whistling a marching tune, they stamped in time to the pulsating structure. The vibrations increased. As the head of the line approached the west side of the bridge, a noise not dissimilar to random gunfire was heard. This was the sound of snapping stay chains and bolts, and one of the iron column supports fell toward the bridge, bringing masonry with it. Around forty of the seventy-four soldiers were thrown into the water, with about half receiving notable injuries. Fortunately, no one was killed, partly because the bridge was not a great height above the river. An investigation into the accident found that a previously weakened bolt had failed as the soldiers amplified the bridge's mechanical resonance. Their conclusions led to a new British army rule, that soldiers must "break step" when crossing bridges. Crossing suspension bridges with random steps means that soldiers no longer initiate resonance disaster through their synchronized marching. London's Albert Bridge, nicknamed "the Trembling Lady," still bears a plaque requiring soldiers to break step, though as the local Chelsea Barracks closed in 2008, the notice is now only of use to sightseeing groups of South Korean Tae Bo tourists.

Arguably the most famous example of resonance disaster is the Tacoma Narrows Bridge collapse of 1940. The writing was seemingly on the wall of this bridge even before it was completed, ever since the Angers Bridge collapse in France, ninety years previously. In that disaster, 483 military personnel and a maid with three children were plunged into the river Maine, and 200 died (including the maid and children). The cause of the collapse was a combination of marching soldiers, rusty cables, and storm-force winds. And such winds were also destined to spoil the party at Tacoma Narrows over Puget Sound in Washington State. During construction between 1938 and 1940, even moderate winds caused the deck to rise and fall with enough significance for the workers to nickname the bridge "Galloping Gertie"—which should have set the designers' alarm bells ringing.

But this wasn't the original bridge design. This was Tacoma Narrows Bridge version 2.0. The first was developed by an engineer named Clark Eldridge but was rejected due to its high cost. Eldridge was partly vindicated—though he did accept some of the blame for the collapse—as the 1950 replacement bridge (version 3) has many similarities to his original design. However, the "budget" Tacoma Narrows Bridge 2.0 did not possess sufficient stiffness, a cost-cutting modification made to Eldridge's original design by engineer Leon Moisseiff. Though Moisseiff's modifications made it cheaper and arguably more elegant, we now know that elegance at the expense of functionality can end up in a heap at the bottom of Puget Sound. Eldridge witnessed the bridge's downfall, an undoubtedly devastating moment for him. Another witness to the collapse had an even more devastating experience. Leonard Coatsworth, a reporter at the *Tacoma News Tribune*, had a very lucky escape.

I drove on the bridge and started across. In the car with me was my daughter's cocker spaniel, Tubby. Before I realized it, the tilt became so violent that I lost control of the car. I jammed on the brakes and got out, only to be thrown on to my face against the curb. Around me I could hear concrete cracking. On my hands and knees most of the time, I crawled 500 yards or more to the towers... Safely back at the toll plaza, I saw the bridge in its final collapse and saw my car plunge into the Narrows.

And yes, Tubby went down with the car and most of the road deck. Coatsworth, along with a few others, had attempted to coax the dog from the car, but to no avail.

The Tacoma Narrows Bridge collapse is always cited as a classic harmonic resonance case study, though its demise was the result of a much more complicated set of vibrations, caused not by soldiers but by wind. Fundamentally, the bridge's design was a mess, allowing the wind to play havoc with its structural integrity. Differing wind directions caused different vibrational wave patterns as well as dangerous vortices that exacerbated an aerodynamic phenomenon known as "flutter." The road deck twisted in certain conditions, with a motion similar to that when one wrings out a wet towel—one half of the bridge would twist one way, the other half in the opposite direction. In addition to this torsional vibrational movement, the bridge also suffered from a transversal vibration mode. A transverse wave goes up and down like the motion of ocean waves or an earthquake's S waves, as if one is shaking the dust out of a long rug. There is plenty of footage of cars on the bridge bobbing up and down on the road deck, appearing on a wave crest then disappearing into a trough. On certain windy days, the bridge suffered from torsional vibration mode. On other days, when the wind blew from another direction, the bridge wobbled with transverse vibrations. And for the adrenaline junkies, if the winds were optimal, there were days when it was subjected to both sets of vibrations simultaneously, and a whole lotta vortex-induced fluttering was going on.

On the day of the collapse, the transverse waves had a frequency of about 0.6 Hz (a low D) and the torsional waves around 0.23 Hz (a lower sharp A). Was this fact possibly the inspiration for the instantly recognizable intro to the Red Hot Chili Peppers' hit "Under the Bridge," the first chord of which spookily features a D and an A, albeit nine octaves higher than Tacoma? Perhaps not, but it takes no great engineering expert to predict that you cannot regularly twist steel up and down and wring it out without there being consequences. Plans were afoot to try to alleviate Galloping Gertie's misery, but too much towel wringing and rug shaking had already occurred, and the winds from a particularly vicious low-pressure system on November 7, 1940, caused the 850-meter span to give up the ghost, plunging its road deck (and Tubby) around 60 meters into the ice-cold waters below. It seems that a combination of penny-pinching, poor design, and an amnesic attitude to suspension bridge history led to this most infamous of engineering disasters.

A Bridge of Sighs

The creators of a more recent wobbling bridge averted disaster by "listening" to their bridge. One of London's commissioned landmarks celebrating the start of this millennium was a pedestrian bridge spanning the Thames, the first for over a hundred years. The "blade of light" was the winning design created by architects Foster and Partners, sculptor Sir Anthony Caro, and engineering group Arup. The Millennium Bridge footbridge was opened to great excitement on June 10, 2000, with an estimated one hundred thousand people crossing it on its first day. But from the very outset, there was a problem. As the number of people on the bridge increased—you guessed it—the bridge started wobbling. This wobble was not one that the engineers were prepared for. It wasn't a vertical motion like that of Broughton Bridge or a torsional or transverse motion like that of Tacoma. This was a very alarming sideways, lateral movement. Just like the Tae Bo Twenty-Three in the Techno Mart tower, the Millennium pedestrians were compounding the problem. As the bridge moved from side to side, so the walkers tried to steady themselves by widening their gait and walking in step with the bridge, which thereby amplified the lateral motion. And as there were thousands on the bridge (not just twenty-three sweaty South Koreans), the situation was quickly heading in the direction of resonance disaster.

Engineers, technical consultants, and project directors witnessed the very perceptible wobble with disbelief, and it didn't help that their failure to anticipate the phenomenon was causing seismic waves all over the media. This brand-new bridge, meant to portray the age of "cool Britannia," was now suggesting "freezing-cold-in-the-waters-of-the-Thames Britannia" if the experts couldn't very quickly figure out how all their exhaustive testing had missed such a fundamental error. The codes of practice and regulations that the architects and engineers adhered to at the time were based on the mistakes and experiences of generations of previous bridge builders. It seemed unimaginable that a newly designed twenty-first-century bridge could behave in such an unpredictable way.

Although every component in the structure had been tested both individually and collectively, the one test that Arup could not do until the opening day was put *thousands* of footsteps on the bridge at the same time. The engineers had studied the vertical forces exerted on a structure caused through walking. But it seems that they hadn't considered the possibility of a lateral feedback loop, where pedestrians unconsciously synchronize their steps in an innate reaction to a laterally swaying bridge, causing the bridge to wobble further, leading the pedestrians to amplify the bridge's motion even more, etc. The Millennium pedestrians were quickly termed "tuned active exciters," in that this mass of people were treading with the same frequency as the swaying bridge. The bridge was shut down after only two days, and subsequent tests revealed that the structure did not have just one resonant frequency: the north span's natural frequency was 1.05 Hz (a slightly sharp C, eight octaves down from middle C), the south span's frequency was 0.77 Hz (a slightly sharp super-low G), and the central span came out at 0.49 Hz (a slightly sharp low B). If the spans' collective resonances were transposed up, between eight and ten octaves, the Millennium Bridge, when packed with pedestrians, would ring out a major seventh chord in C, similar to the chord at the start of Eric Satie's "Gymnopédie no. 1."

If the engineers of the Millennium Bridge could tune its resonant frequencies above the upper limit of walking (around 1.3 Hz), they could eliminate the chance that it would become "excited" enough to synchronize everyone's steps. The only way to achieve this was through stiffening the bridge, just as Clark Eldridge had originally planned for his span across Tacoma Narrows. But in the case of the Millennium Bridge, such stiffening would compromise the aesthetic design of the "blade of light," meaning considerable structural changes, costing time and money. Instead of trying to raise the frequency of the bridge, they opted to damp the vibrations. This meant fitting over eighty dampers, both linear viscous and tuned mass dampers. Tuned mass dampers are fitted to many of the world's tallest buildings, the most famous being the open pendulum structure near the top of the Taipei 101 skyscraper in Taiwan. In essence, the free-swinging pendulum's motion (weighing nearly eight hundred tons in the case of the Taipei 101) compensates for the movement of the structure it is attached to, as is the case in the Techno Mart tower in Seoul. It absorbs the energy of the swaying building, or bridge in the case of the Millennium Bridge. And just like an old-fashioned analog metronome, the frequency of the pendulum can be adjusted by moving its weight, thereby changing the length of the pendulum. Using this technique, the pendulum's frequency can be matched to that of the structure, maximizing its energy absorption rate.

As well as tuned mass dampers, the engineers fitted viscous dampers, giant syringes full of very gloopy gunk that also absorb the energy from the bridge's vibrations. Again, these can be tuned, using differing levels of viscosity to maximize their effectiveness. And to be on the safe side, the bridge was additionally equipped with new dampers that not only reduced lateral vibrations but also helped minimize any further Tacoma-style vertical or torsional vibrations that could potentially cause problems. Like all structures, suspension bridges are designed to move, but not as much as Broughton, Tacoma Narrows, or the Millennium Bridge did. On February 22, 2002, Londoners could no longer teasingly call it the "Wobbly Bridge"—the dampers did the trick, and the bridge reopened to the public. This high-tech collection of syringes and pendulums foiled the thousands of "tuned active

exciters" (usually known as pedestrians), and the bridge has suffered no further wild vibrations. But has anyone held a Tae Bo
workout session on it?

GHOST NOTES

The mysterious world of infrasound

Late one dark, dark night in the early 1980s, a laboratory at Coventry University, England, became the setting for a chilling ghost story of spine-tingling dread—featuring a scientific sleuth, a deathly gray apparition, a fencing sword, an eyeball, and a petrified janitor...you get the picture. Engineer and head of the lab Vic Tandy was a typically logical, levelheaded scientist, but even he could not deny that, at this late hour of a dark night in his lonely lab, he was seeing a moving gray apparition out of the corner of his eye. Bathed in a cold sweat, his heart pounding in his chest, he questioned his own rational mind—could he be having a supernatural encounter?

Tandy's lab was not particularly gothic; it was a couple of garages joined end to end, about three meters wide by ten meters long. His work centered around designs for new medical life-support equipment. For some time, the janitor had warned Tandy of distressing sightings, ghostly goings-on, and strange feelings of dread and depression in the lab. A hard-nosed engineer like Tandy was not going to fall for this. But it wasn't just the janitor. Other members of his team reported occasional feelings of unease and experiences of doom-laden depression. On one occasion, a colleague turned to speak to Tandy, convinced he was standing next to him...but Tandy wasn't there. He was at the other end of the room.

On the evening in question, long after his colleagues had gone home, Tandy began to feel that same sense of anxious dread himself. He started shivering and was convinced that he was being watched. He claimed that a gray, blob-like figure appeared in his peripheral vision and moved silently from his left. Tandy said it moved as a human would but was completely indistinct. He was terrified, the hairs on the back of his neck standing to attention. Eventually, he plucked up enough courage to turn his head and face the apparition, but as he did so, it faded and then vanished. Had Tandy been working too long and allowed fatigue to get the better of him? Had he lost control of his rational thoughts, leaving his imagination to run riot? Or was it just possible that this laboratory was truly haunted?

Returning the following day—undoubtedly with a mix of trepidation and curiosity—Vic Tandy sat down at his desk with, of all things, his trusty sword. This wasn't, however, meant to be the solution to the problem. Tandy was a keen fencer and was preparing his blade for a forthcoming competition. He clamped it in a vice and went off to collect various pieces of sword-preparing equipment. When he returned, the free end of the blade was vibrating wildly. The fear he felt the previous evening immediately returned, but then, after some thought, it seemed unlikely that the cold metal of a fencing sword would be the material through which the dead might communicate with the living. There were no discernible drafts, loud sounds, or earthquakes in progress, so why was his sword swaying?

Tandy set to some good old-fashioned ghost busting, using his considerable scientific and engineering knowledge. He guessed that his blade was probably vibrating in sympathy with a frequency outside human audible range. The lowest note on a standard piano is A_0 at a frequency 27.5 Hz. Beneath this, the Infinite Piano only has about six more playable white keys before our ears begin to fail to discern individual pitches. As we dive down past 20 Hz, we're in that mysterious world of *infrasound*.

As any good engineer would, Tandy set about experimenting. He placed his sword in a portable vice and moved it around the room. Curiously, the blade swayed most vigorously in the middle of the room. As he moved toward the extremities of his lab, the vibrations subsided. Tandy suspected his sword was being stimulated by a standing wave.

Picture Jell-O with a cherry placed atop its center. For the purpose of this analogy, the Jell-O is the air filling Tandy's lab, the cherry is his sword, and a pesky child prodding the side of the Jell-O replicates the infrasonic sound source. Directly after the child's first prod, the Jell-O lunges forward. If there were no wall at the end of Tandy's lab, the Jell-O would spring back naturally. But because of this obstruction, the Jell-O springs back farther as it also contains the wave energy that is reflected back off the wall's surface. The Jell-O is now prodded again at the precise moment that the reflected wave bounces off the wall from where the prod comes. Consequently, the Jell-O lunges even farther forward than before, due to the combination of the new prod *and* the reflected wave. Looking from above, the cherry is swaying back and forth with more energy than the initial push could have generated. This wild wobbling cherry is what Tandy saw in his sword.

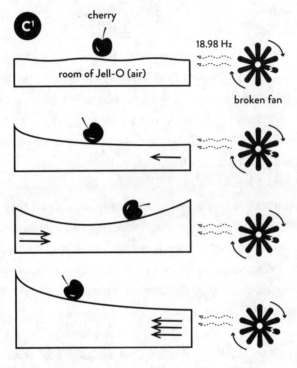

A Standing Wave, Cherry, and Jell-O

Tandy measured the length of the room and, using the velocity of sound, calculated a frequency for his standing wave (in Jell-O terms, how many times per second it is prodded). It was 18.98 Hz, around the musical pitch D_0, forty-seven white notes to the left of a piano's middle C, or four notes off the low end of a standard piano. But what or who was producing 18.98 Hz, a frequency on the cusp of human hearing?

Rather disappointingly, and not in any way spine-chilling, the culprit turned out to be a newly fitted fan in the extraction system. When the fan was turned off, the sword lay still and the standing wave disappeared. But perhaps of more interest was how Tandy and his team had reacted psychologically to the infrasonic standing wave. What was it in this frequency that gave them all cold sweats, ghostly visions, and a sense of doom? For many, working in a garage in Coventry would be depressing and doom-laden enough, but surely it wouldn't drive them to see dead people.

Seeing Stars

The answer seemed to come from research commissioned by NASA, which had instructed scientists to investigate this very question (the effect of infrasound on human physiology and psychology, that is, not the effects of working in a garage in Coventry). In a 1976 report, the Aerospace Medical Research Laboratory hypothesized that around 19 Hz could be the resonant frequency of the human eyeball. Just as in the famous opera-singer-and-the-wineglass trick, at this frequency, NASA suggested that eyeballs vibrate rather alarmingly. If you were exposed to 19 Hz while reading this book, the words on the page would become smeared and your vision impaired. Such a visual impairment could explain the blob-like figure that Tandy had seen. Another space-led report found that infrasonic vibration causes hyperventilation, which is often associated with panic attacks, muscle cramps, breathlessness, and a sense of impending doom and death.

But what sparked NASA's interest in the connection between infrasound and its effect on humans? Within only sixty years of the Wright brothers' first ever heavier-than-air controlled flight in 1903, men and women were flying in outer space. There was one major weakness during this phenomenal rate of progress—the fragile human body. The relative slow burn of human evolution was not keeping pace with the rapid developments in aeronautical technology. During flights, astronauts and high-altitude pilots struggled to breathe, stay in their seats, or indeed remain conscious. Engineers started to build comprehensive research programs to study the physical effects of things like speed, altitude, g-force, and vibration. A number of facilities were tasked with putting pilots through very unpleasant experiences to ascertain how much physical and mental stress the human body could withstand.

Langley Research Center in Virginia was one such establishment. Machines, test pilots, and future astronauts were shaken and stirred, rocked and rolled, until they could stand no more. The lessons learned were invaluable in exploring how humans function when flying at extreme speed and very high altitude. Langley Research Center housed the Low-Frequency Noise Facility, a room made for large-scale acoustic tests of both machines and humans. As its name suggests, the center specialized in researching low-frequency noise and vibration, from 50 Hz (around the first funky bass note of Herbie Hancock's "Chameleon") downward into infrasound. The room was built to replicate the extreme intensity of infrasonic vibrations that a space capsule and its inhabitants are subjected to during the launch of a multimillion-pound thrust rocket. A 1966 report found that, after blasting subjects with short bursts of high-decibel infrasound, they experienced feelings of annoyance and discomfort, and fatigue slowed their performance rate. Mission Control could work with annoyed and discomforted astronauts, but fatigued and slow-performing ones could be dangerous. Pogo oscillation, as NASA calls such vibrations, has occurred on many space missions, notably in 1965 on Gemini 5, when astronauts Gordon Cooper and Pete Conrad had their vision and speech momentarily impaired by the strong vibrations of their shuddering rocket. Other research noted that below 25 Hz (the first G♯ off the bottom end of a standard piano, or the lower frequency that some cats purr at), astronauts experienced some modulations of speech and an annoying, full-throated gag sensation. Have a pleasant flight, everyone!

Since those early space missions, scientists have continued to improve their understanding of the relationship between human health and infrasound. If 19 Hz sets one's eyeballs vibrating wildly, could other parts of the human body be susceptible to specific frequencies, not just in the infrasonic range but right throughout the frequency spectrum? Researchers have used dummies, animals, and even dead humans to discover whether specific frequencies make parts of us wobble uncontrollably. Because human bodies vary so widely, results have never been completely reliable. However, some recent NASA research suggests the following: our pelvic area naturally resonates around 8 Hz, our hands between 1 and 3 Hz and 30 and 40 Hz, and our chest wall about 60 Hz. Our whole body standing erect is between 6 Hz and 11–12 Hz, and lying down is 2 Hz. Based on this, the chord that seems to resonate notable sections of our bodies (when sitting down) is A♭maj^{7}♯$^{9(add2)}$/C, though at such low frequencies, we wouldn't hear most of it—except for the resonant frequency of our eardrums at 1,000 Hz. Transposed up five octaves into the piano range, the body chord of A♭maj^{7}♯$^{9(add2)}$/C looks very impressive though sounds less so.

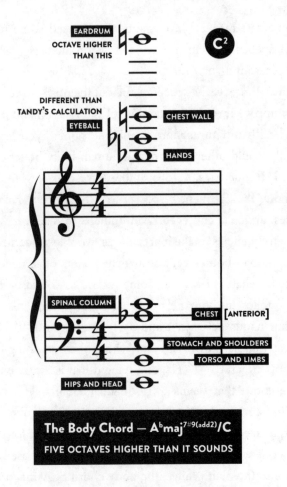

The Body Chord — A♭maj⁷#⁹⁽ᵃᵈᵈ²⁾/C
FIVE OCTAVES HIGHER THAN IT SOUNDS

Back in spooky Coventry, Vic Tandy managed to retain an open mind when it came to the paranormal. He eventually became, rather surprisingly, a fully signed-up member of the Society for Psychical Research, saying that "when it comes to supernatural phenomena, I'm sitting on the fence. That's where scientists should be until we've proved that there isn't anything." The human imagination has always been fired by the spirit world, and Tandy was a member of an organization that had weathered many a skeptical storm. The society was founded in 1882, a time of fervent interest in spiritualism, mysticism, and the occult. In those days, the society was taken very seriously, counting among its members British prime minister Arthur Balfour and writer Sir Arthur Conan Doyle. Indeed, throughout the late nineteenth and early twentieth centuries, many people would have believed and fully embraced a paranormal explanation of Tandy's encounter.

It's All About the Bass

The relationship between the occult and infrasound, as well as much of the lower end of the audible frequency range, has historically been a strong one. During the time of the founding of the Society for Psychical Research—the late Romantic period, as cultural historians like to call it—composers of spooky music dragged melodies down into the bowels of musical pitch, areas that until then were reserved for rather boring bass lines. They plundered the lower end of the frequency range in order to enhance the expressive qualities of their music. For much of the previous era of classical music—handily called the Classical period (about 1750 to 1825)—the likes of Mozart and Haydn stuck to the established rules of high melodies and low bass parts. Melodies were written for the upper instruments such as flutes, oboes, and first violins, tedious chugging chords were played by the likes of second violins and violas, and equally boring and functional bass parts were left to the cellos, basses, and bassoons. But as the nineteenth century progressed and composers looked to new conventions to express their increasingly fervid imaginations, bass instruments started to feature more and more in melodic writing. The use of low, rumbling bass tunes litters the world of mystical and doom-laden orchestral music.

Classical Era

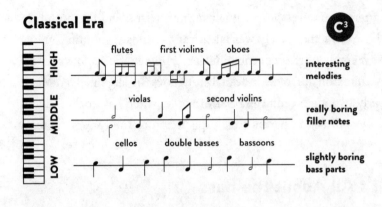

Scary Romantic Era

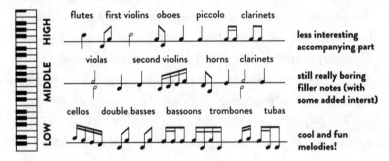

One of the most famous orchestral works of the era tells the story of a drug-induced murderous crime of passion, which results in a beheading and a tormented gathering of demons and sorcerers. Hector Berlioz, himself an opium addict, wrote the *Symphonie fantastique* in a frantic six-week period during 1830. It was way ahead of its time. The last movement—"Songe d'une nuit du sabbat" or "Dream of the Night of the Sabbath"—features an ancient plainchant melody, the Dies Irae, the Day of Judgment theme. As if that wasn't scary enough, it is played on one of the lowest and loudest instruments of the orchestra, the tuba—two of them, in fact, just to double the scare factor. Berlioz wanted to explore the orchestral depths to intensify the ghoulish and supernatural fervor of the movement. At the opening of the piece, just after the creepy footsteps in the trombones, the first tuba sinks to a bottom C around 32 Hz. (Many modern professional players can easily get an octave lower, down to a super low C at around 16 Hz, and even beyond.)

A few decades later, the Russian composer Modest Mussorgsky was exploring his own penchant for musical spookiness. *Night on the Bare Mountain* is perhaps Mussorgsky's best known work, a terrifying orchestral tour de force inspired by a short story in which St. John witnesses a witches' Sabbath on a mountain near Kiev. The mountain is awash with trolls and all things evil that come out to dance and frolic at midnight. After a swirling introduction followed by hammering violins and woodwind, Mussorgsky gathers a host of demonic bass instruments to blast out a petrifying melody, which ends with a plunging leap of doom toward the climax of the phrase. Bassoons, trombones, tuba, violas, cellos, and double basses combine, all exploiting their lower ranges to express the spine-chilling atmosphere of the mountain, the tuba blasting out a 25 Hz low G♯ on the stabs at the end of the first phrase.

A century before this work, composers were unable to imagine the potential emotional power of turning an orchestra upside down, of placing extended melodies low in the frequency range and leaving all the accompanying parts above the main action. The fashion then wasn't about expressing emotion but about structure, and as the trend for ordered restraint waned, the shackles of creative convention were cast off, and all kinds of new techniques emerged. Inverting the orchestra and treating listeners to full-volume low frequencies must have been as new and exciting for the listeners as it was for the composers. Other works, such as Edvard Grieg's "In the Hall of the Mountain King" and Paul

Dukas's bassoon-heavy *The Sorcerer's Apprentice* also top the mystical Romantic low-frequency chart. Although none of these melodies are quite in the infrasonic range, their powerful bass notes had a similar emotional effect on the nineteenth-century listener as the extractor fan's infrasonic wave had on Vic Tandy—the frequencies shook the body and stirred the mind, daring the imagination to conjure dark netherworlds of spirits, death, and hell.

For Tandy, his ghostly sword-fencing experience had proved to be an amazing set of coincidences: the extractor fan's wavelength was exactly twice that of the room, thereby creating a standing wave and sweet spot (antinode) of acoustic energy in the middle; his fencing sword reacted to this standing wave; its frequency was at almost 19 Hz, meaning that the apparition he saw was possibly caused by his eyeballs' resonance; and Tandy was a curious engineer who had the scientific know-how to work through the problem.

Subterranean Singers

What NASA and Vic Tandy "discovered" may not actually be so new. In fact, it may be a *rediscovery* of something our ancient predecessors understood. In recent years, a number of people have crawled into caves and prehistoric human-made structures on the hunch that our ancestors might have been more attuned to frequency than we probably give them credit for.

Camster Round in the Highlands of Scotland is a passage grave, one of two large Neolithic chambered cairns at the site. Its underground design is bottle shaped, with a passage leading to a burial chamber. Researchers from Reading University in the UK calculated that this shape has a resonant frequency deep in the infrasonic range, between 4 and 5 Hz. Unless Neolithic humans had very different ears from us, they would not have been able to hear this frequency. Undeterred by this, the researchers took a drum and an audience to the site. Though the drumming sounded distorted at various places in the cairn, its sound traveled no farther than 100 meters away from the outer perimeter. However, rather bizarrely, listeners in another monument 190 meters away could perceive the beat. It apparently sounded "as a distant 'booming' which appeared to rise from the ground." In tests using vocalizations, "listeners occasionally perceived these sounds to be contained within their heads, which could be unpleasant." Supported by other experts' research, these findings suggest it is more than likely that Stone Age people were aware of the acoustic properties of their surroundings.

With the help of a trained singer, leading acoustic archaeologist Professor Iegor Reznikoff identified specific places within inhabited French caves where audible resonance was most notable. He discovered that acoustic sweet spots—those places where a vocal sound seemed particularly to bloom, resonate, and reverberate—were often the same places where cave paintings were found. Indeed, in some more inaccessible tunnels, Reznikoff discovered little painted red dots where the human voice was especially resonant and/or amplified. At various Stone Age locations in the UK and Ireland, archaeoacoustic expert Paul Devereux and Professor Robert Jahn generated tones across the frequency range and quickly established that certain frequencies produce standing waves, just like the one experienced by Vic Tandy. The burial chambers, long barrows, and cairns help to amplify such frequencies. But what was particularly startling— and more than a little coincidental—was that from the cairns of County Meath in Ireland to the megalithic tomb of Chun Quoit near St. Ives in Cornwall, many of the structures reveal resonant frequencies between 95 and 112 Hz. Indeed, in conversation, Paul Devereux told me that he thought many of the tombs had been "retrofitted and almost engineered" with rocks in order to enhance the resonance in this frequency range.

In 2008, a team of psychiatrists, psychologists, and neuroscientists drew on Devereux's findings to reveal that brain function changed at around 110 Hz, stating that "activity over the prefrontal cortex shifted from one of higher activity on the left at most frequencies to right-sided dominance at 110 Hz. These findings are compatible with relative deactivation of language centers...that may be related to emotional processing." In other words, a blooming resonance around the note A_2 in an enclosed space might just push our natural emotional buttons. (A_2 at 110 Hz is the note Leonard Cohen descends to on the word "bird" at the start of "Bird on the Wire.") Such low frequency singing within a small frequency range is most often associated with chanting, where the range of musical pitches is limited to just a few notes, and this is probably no coincidence. Today, many ancient cultures—in countries from Mongolia to Canada—have preserved chanting traditions, among which is the practice of throat or overtone singing. It requires a great deal of vocal skill, as the singer is essentially producing two notes simultaneously. It often has a very limited vocal range and consequently is perfectly suited to the narrow limits of the cairns' resonant frequencies. A skilled chanter could easily, with a little practice and a good ear, tune the pitch of their chant to resonate and bloom within the walls of these tombs.

One particular example of ancient low-frequency chanting is that of Tuvan throat singing, which remains popular in Inner Mongolia and parts of Siberia (as well as with Sheldon in *The Big Bang Theory*!). The sound of a Tuvan throat singer is not unlike a duet between a didgeridoo and a small whistle pipe, except it's sung by one person. The vocalist emits a low drone note, which is then manipulated by both the folds in their vocal cords and their mouth to produce overtones. Overtones are a series of other frequencies that are blended within a given pitch, like the colors of a rainbow within white light. All musical notes produce overtones, and the differing mixture of overtones is what allows us to tell, for example, the difference between instruments' timbres or sound qualities.

There is a mathematical formula found throughout nature that all overtones follow when a musical pitch—sometimes called the fundamental—is sounded. Think of it a bit like a recipe. A beef Bolognese contains beef as its fundamental; tomatoes, onions, garlic, red wine, and herbs are its overtones. Add an extra glass of red wine and your guests will love its richness and depth. However, for a children's version, it's probably best to reduce the wine and garlic drastically and increase the tomatoes. This is what the Tuvan throat singer is constantly doing within a split second, altering the overtone mixture to create a whistle-like melody above the drone. And profoundly impressive it is too.

Back at the cairn, imagine the effect of a low-frequency chant at a burial or a spiritual gathering, where the singer's voice seems to grow in volume beyond its natural dynamic, an almost otherworldly voice amplified, accompanied, and distorted by the gods and spirits. Could it be that our ancestors exploited frequency to enhance their spiritual rituals and beliefs? The acoustic relationship to cave paintings in France and the discovery of consistent resonant frequencies in many UK and Irish cairns gives us a more than tantalizing suggestion that Stone Age civilizations understood something of frequency and how it could be used in music to further intensify spiritual events.

Pipe Down

In the centuries since, infrasound has continued to play a part in our spiritual lives. From about 1500 onward, the pipe organ began to assert itself as *the* sacred instrument. The organ literally became part of the furniture in cathedrals, churches, and chapels. During its development, there was a drive to increase the range of the instrument, not only in terms of its scope of timbral colors through its stops but also in its range of frequencies. There are, of course, physical limitations to how low a pipe organ can play; the lower the note, the larger the pipe needed (that's why a tuba is bigger than a trumpet). One must actually raise the roof to accommodate longer pipes that can play lower notes. To drop by an octave on a pipe, it must be twice as long. (To get a sense of an octave leap, listen to the end of the opening of Jeff Wayne's Musical Version of "War of the Worlds," in those weird synth "weeooos" followed by a heartbeat-like bass drum.) Over the centuries, church builders and ecclesiastic budget holders must surely have questioned whether it was worth constructing and installing pipes large enough to produce notes on the cusp of human hearing. It seems that organ builders were on to something, though, as there is now a suggestion that there can be spiritual benefits to these low frequencies.

In 2003, composer Sarah Angliss gathered a team of psychologists. She was intrigued by reports that infrasound affects people both physically and emotionally, and she wanted to scrutinize these claims scientifically in a musical setting. Many have claimed that infrasound created by low organ pipes intensifies religious feelings, especially during a monstrous final chord at the end of a hymn or anthem. Angliss was also curious about Vic Tandy's experience with his ghostly extractor fan. Her piece for piano and electronics, titled *She Goes Back under Water*, was performed at the Purcell Room in London, where, to expose the audience to infrasonic frequencies, the team had to build a special generator, as most standard musical equipment isn't manufactured to accommodate anything under 20 Hz. To mask the infrasonic notes from her unknowing audience, Angliss blended them with a range of very low audible frequencies. The concert was performed twice, once with the infrasound and once without. Data from the blind test concerts (not even the infrasonic generator operator knew which concert featured the infrasound) was collected by way of audience questionnaires, asking about their feelings throughout both performances. Twenty-two percent of the audience admitted to more "unusual experiences" during the infrasonic performance, including odd stomach feelings, shivering, anxiety, changes of heart rate, and flashbacks of emotional loss. Although perhaps not completely conclusive, Angliss's experiment seems to support what many people have experienced over centuries. When the giant pipes are played at high volume, the sensation can be more than just aural—members of congregations claim to "feel" different, their bodies and minds responding to a holistic sensation, one they can only explain as a moment of heightened religious awareness.

Perhaps the most famous of all organ works, the Toccata and Fugue in D Minor by J. S. Bach (the dramatic opening music to Disney's iconic movie *Fantasia*), is one of those pieces that can make a serious impact on the structural integrity of both congregation and church building. And one doesn't have to wait long before infrasonic frequencies start appearing. After the first three descending flourishes, the thundering low D pedal note is then built on with a calamitous diminished seventh chord before resolving through a suspension on to a dark D minor chord. Through a funny quirk of physics, combinations of notes (which are clearly in our hearing range) can also create infrasonic by-products, so the two chords at the end of this section, along with the low pedal D, all have the potential, on a large pipe organ, to send the listener into religious fervor. Similarly, the ending of César Franck's Choral no. 3 has bone-shaking infrasonic potential. The last section features a slow-moving bass line under some fiendishly difficult chromatic writhings and ends with two chords that form a perfect cadence (in layman's terms, a musical period). The penultimate pedal note is written as D_2 (73.42 Hz), but with a rumble-loving organist pulling out the right stops, this would sound at D_1 (36.71 Hz), and with big enough pipes, potentially D_0 (18.35 Hz), delving into the eyeball-resonating infrasonic zone. The choice of stops for this particular piece (its organistic term is "registration") was the very last musical activity Franck under-took in October 1890. He had been hit in the chest by an omnibus the previous May, and a chill in early autumn turned quickly to pleurisy and eventual death. His last action was to struggle up to the organ loft at the Basilica of Saint Clotilde in Saint-Germain-des-Prés in Paris to choose the right registration for the chorals. It's tempting to wonder whether the infrasonic frequencies of that final perfect cadence even accelerated their composer's own demise.

And into the twentieth century, another Parisian organist and composer continued to expand the potential of the instru-ment. Olivier Messiaen was organist at Église de la Sainte-Trinité for over sixty years. His use of extended theoretical musical tech-niques is now regarded as one of his trademarks, and Messiaen loved to combine musical theory with his devout Roman Catholic faith. La nativité du seigneur (The Birth of the Lord) is regarded as one of Messiaen's early masterworks. It is divided into nine movements called "meditations," each inspired by the birth of

Jesus. From an infrasonic point of view, the final meditation is of most interest. It is titled "Dieu parmi nous" ("God among Us"), and in its shockingly loud dynamic at the start (especially following a very quiet eighth meditation), one can sense that Messiaen really does want the listener to feel the immediacy and intensity of a godly presence in the church. The opening two short downward patterns are marked *fff* (that is, very loud indeed). But when the extremely slow descending pedal part enters, Messiaen adds another *f* for good measure and starts the plunge into the depths beyond the D_2 in Franck's Choral no. 3 to C_2 (65.41 Hz). As previously explained, with a few extra stops, the note can drop an octave to C_1 (32.7 Hz) and possibly down again to C_0 (a whopping 16.35 Hz), arguably a note that our ears cannot really hear but one that feet on floors and bottoms on pews certainly can feel. With such loud infrasonic waves resonating around a church, "God among Us" is not beyond the realms of possibility. So as both church congregations and NASA's astronauts gaze heavenward, they share a common experience: the profound physiological effects of human-made infrasound.

Antisocial Bubbles

It is not just explorers looking to the heavens that need to be aware of the effects of infrasound. Divers who journey down into the oceans' depths must also beware of inaudible rumbles and their potential ill effects. Sound traveling through water behaves very differently from the way it behaves in air. High frequencies are quickly attenuated (absorbed), but low-frequency waves travel faster and farther in water, and infrasonic waves can have a devastating effect on the submerged human body. In water, a diver's lung is effectively a bubble, a large pocket of air surrounded by liquid. Bubbles are very odd, wildly dynamic structures that infrasound has a considerable impact on. Dutch astronomer Marcel Minnaert is best known for his work on stellar atmospheres and spectroscopy, but he was also fascinated by bubbles. His 1933 article "On Musical Air-Bubbles and the Sound of Running Water" described the structure of bubbles, how they act in liquid, and how the combined sound of billions of these makes the "musical" sound of running water. Minnaert devised an equation to calculate the resonant frequency of a single bubble—it is aptly called the Minnaert resonance.

The gas in a diver's body, often in bubble form, is an area of serious concern. Many military tests have attempted to study the effect of sound waves on a submerged human. One of the main underwater threats is blast waves from explosions, and much research has focused on what impact these have. In the 1940s and 1950s, the UK's Royal Naval Physiological Laboratory carried out extensive underwater blast tests. Their research began with animals, but the laboratory moved in 1949 to large-scale, live human testing. Those subjected to the blasts suffered a range of symptoms including chest pains, paralysis, concussion, and aural damage. Using a specially designed formula—undoubtedly with a little help from the Minnaert resonance—the report suggested that a 150-pound person would have a lung resonance of around 45 Hz. (It so happens the first two notes of the *Jaws* theme are between 41 and 44 Hz.) Body tissue has a similar density to water, so sound waves do not greatly affect a diver's body; they pass through tissue. However, the gases inside a body are more compressible and therefore susceptible to wave pressure fluctuations, leaving the interfaces between body tissue and gas very vulnerable. Blasts particularly affect lungs, sinuses, and ears. Air spaces in the head are also at high risk when submerged. A more recent report's list of possible underwater blast symptoms includes edema, severe hypoxia, dyspnea, tachypnea, cyanosis, hemoptysis, pulmonary disruption and hemorrhage, arterial gas embolism, respiratory failure, transient paralysis, testicular pain, nausea, vomiting, and an urge to defecate...all assuming that Jaws doesn't get you first. Models predict that the greatest amount of tissue strain with low-frequency sound occurs in the trachea and lungs between 30 and 40 Hz. If you have ever considered repeatedly playing the first bass note of Wild Cherry's "Play That Funky Music" (around 40 Hz) on a bass guitar plugged into a gigantic amplifier while in a swimming pool to test the limits of your central airway, you might want to think again.

Throughout our lives, most of us are completely unaware that our bodies are shaken and wobbled by a vast range of both audible and inaudible frequencies. Some researchers now suggest the eyeball's resonant frequency is more likely to be between 60 and 90 Hz (in the middle of this range lies the D_2 played by the cello at the start of the tear-jerking third movement of Elgar's Cello Concerto) and not 19 Hz as NASA and Vic Tandy were led to believe. However, what we *do* know is, to be on the safe side when performing on our Infinite Piano, it might be best to avoid repeatedly striking the A (19-ish Hz) that lies forty-seven white keys to the left of middle C. Both the performer and the audience could break out in cold sweats, get incredibly depressed, and might even start seeing dead people.

FIX IN THE MIX

Chaos unleashed when blending frequencies

As compelling as the stories of individual frequencies are, the combination of two or more pitches opens immeasurable worlds of complexity and wonder. It is such combinations that give music its emotional intensity, its color, its life. This mixing and blending of frequencies is the perfect collaboration of art and science. The full vibrational complexity of the opening E♭ major chord of Beethoven's *Eroica* Symphony, performed by a professional orchestra in a large concert hall, contains enough physics to fill this whole book. During its explosive first chord, each cellist in the orchestra plays a single E♭ (all at slightly different frequencies roughly around 155–156 Hz...hopefully), resonating their own instruments and consequently their own bodies. But they are also affecting the other cellos and cellists near them, creating a collective cello sound that excites the air around the whole section (and the double basses and woodwind instruments in the vicinity), generating sound waves in that particular part of the auditorium that behave differently according to whether the seats are occupied, the floor is carpeted, or the exit doors are open or shut. And we haven't even started exploring the series of overtones hidden within the cellos' E♭s, which resonate other instruments and parts thereof, the music stands, the wood panels, or the ushers' dental fillings. And then consider that this is just the cello section...and only its first note. There are another fifty minutes of a mind-blowing multiplicity of combined vibrational convolutions to get through. It's hardly any wonder that we let this frequential cocktail simply wash over us, allowing our hearts to wobble more than our minds.

But let's put aside our seventy-piece symphony orchestra blending frequencies in a chord and explore the less complicated combination of just two frequencies. What makes some pairs of pitches sound appealing while others sound dissonant? The answer lies in ratios, the subject that Johannes Kepler got a bit hung up about in chapter A. At its heart, the mathematically simpler the ratio between two given frequencies, the more harmonious the blend. For example, the distance between each consecutive string on a violin is the interval of a fifth—its lowest string is G_3 (195.9 Hz), its next a D_4 (293.6 Hz). The ratio between these two frequencies (294:196 Hz) is 3:2. A_4 (440 Hz) is the next string up from the D, again a ratio of 3:2, and so on. Fifth intervals "blend" when played simultaneously. Indeed, they blend so well that they're everywhere, from the drone of bagpipes to the power chords of heavy metal music. But the success of this ratio, the fifth interval, is not human-made; it can be found throughout nature. (For example, increase the speed of rotation of a whirly tube from its lowest tone, and the note will leap up by a fifth interval.)

Another fundamental interval of music is the major third, the fuzzy, warm-glow combination of frequencies that gives our music its positive emotional vibe. (It's no accident that the *bing-bong* of a classic doorbell is this interval, a "Hello-welcome-to-our-lovely-house-we'll-be-with-you-any-second-now-with-a-beaming-smile-and-a-wafted-smell-of-freshly-baked-bread" musical greeting, neatly condensed into two frequencies.) The major third leap of the first two notes of "When the Saints Go Marching In"—let's say a C (261.6 Hz) to an E (329.6 Hz)—has a ratio of 5:4, but they still have the same ratio if I sing the melody starting on a higher pitch, perhaps from G (391.9 Hz) to B (493.8 Hz).

Intervals that sound less consonant have ratios that are mathematically less "pure." Eventually, we arrive at the devil's interval, the dreaded tritone with a ratio of 45:32 or 64:45 (don't ask why there are two ratios). This *diabolus in musica* was treated with great caution and a very long stick during the Renaissance but was tamed enough (in reality, audiences simply learned to love its apparent "devilment") to become a cornerstone interval from the eighteenth century onward. Perhaps two of its most famous musical examples are from the twentieth century: the opening two notes of Leonard Bernstein's song "Maria" from *West Side Story* (the tritone litters this musical), and the combination of two entangled *diaboli* that form an anxiety-ridden diminished seventh chord, that classic trembling piano accompaniment to the clichéd girl-tied-to-a-train-track-in-front-of-an-oncoming-steam-train black-and-white movie scene.

A similarly dubious interval is that of the semitone, the distance between any two adjacent notes on the piano, black or white. This uneasy clash, most recognizable when heard as two separate notes at the start of "The Pink Panther Theme" (indeed, the first five pairs of notes are semitones), has a ratio of 16:15. Played together, this very dissonant clash of two notes is the smallest interval on a piano. Obviously, though, on many instruments and in many cultures, even smaller intervals are easily achieved through tiny finger movements on a violin or the bending of notes on a sitar, by changing air pressure on a trumpet or clarinet, or simply altering the pitch of one's voice. And what happens as two notes get closer and closer in pitch is an unfolding story concerning baroque violinist Giuseppe Tartini, Victorian police whistles, a round-the-world endurance flight, super scientist Dr. Larry Fogel, and the art of piano tuning.

Rhythm Helps You Tune Hertz Measurements

Contrary to popular belief and seeming common sense, many piano tuners do not rely on the *pitch* of intervals to tune piano strings—they use *rhythm*. For midrange and higher pitches on a piano, an individual note is made up of three strings struck simultaneously, and these strings must be perfectly in tune with each other. To achieve this, piano tuners do not whip out an electronic tuner or open an app on their mobile phones. Bizarrely, they listen to a steady, pulsing beat that changes its tempo depending on how in tune a pair of strings sounds. If one string is vibrating at 440 Hz and another at 445 Hz, when combined, they wobble at roughly the same frequency, but there is a discrepancy between them of 5 Hz. It is possible to detect this discrepancy as either a rise in volume (their combined wave crests increase the amplitude) or a slight decrease as one wave negatively cancels out some of the other wave's energy. At 5 Hz, this happens five times every second, a pulsing change of volume that skilled piano tuners are trained to hear. With a little releasing tweak of the string's tension, 445 Hz might drop to 443 Hz. The simultaneous sounding of the two strings now produces three pulses or "beats" per second, the difference between 440Hz and 443Hz. Another tweak might bring the string down to 441Hz, at which point the beat would sound once every second. By reducing the speed of the beats, piano tuners know they are approaching their target. Of course, go too far the other way and flatten the string below 440 Hz, and the beats start again. Without a need for a flashy box of techno tricks, the piano tuner knows when two strings are perfectly in tune by gettin' down to the beat.

Tuning the three strings that make up the sound of an A_4 at 440 Hz is only half the story...actually, it's the easy bit at the end of the story. Before any of this happens, the piano tuner tunes just one string per note, tweaking all eighty-eight notes to fit with one another. And this is where our modern, sophisticated, swanky tuning system called equal temperament reveals itself as a complete dog's breakfast. Piano tuners actually *detune* pianos to get them to sound "in tune"—it's all a messy compromise. The reason for this is the tension between mathematics, nature, and our own sensibilities.

History is littered with attempts to divide frequency into satisfying musical scales. To illustrate the challenge (through a rather improbable analogy), let us travel to ancient Greece; to a time when the "Ancient Athens Tourist Board" commissioned one of the great minds of the day to design a new hotel. That mind belonged to Pythagoras, and the brief was as follows:

> Up on that hill over there, build us a never-ending cylindrical tower with an internal spiral staircase; twelve rooms per floor (each of which must be proportionally equidistant along said staircase); each room directly above its corresponding one on the floor below but at twice the altitude; the whole internal

structure appearing to follow the shape of an infinitely stretched spring that is attached to the ground...but if it's too difficult, we can ask Archimedes if he's free.

Pythagoras took the gig, though he was a bit nervous, as the only tools he had to help him were an altimeter and his understanding of the ratio 3:2. He decided to name his hotel "The Pythagorean Tuning" (admittedly a strange name for a hotel), the floors "octaves," and the rooms "notes" (A, A♯, B, C, C♯, D, etc.). He immediately set to work on building the first floor of his hilltop-infinite-cylindrical-spiral hotel, which, according to his altimeter, was exactly 55 meters above sea level. Looking directly north, he marked out his first room, A_1. Immediately, he knew that room A_2 would need to be directly above it at 110 meters, room A_3 above that at 220 meters, and room A_4 above that at 440 meters. But how was he going to subdivide a rising and expanding spiral staircase to fit eleven more rooms per floor, all proportionally equidistant? Being clever, Pythagoras knew that the altitude of room E—the seventh room above—needed to be at a ratio of 3:2 compared to room A (he had noticed this relationship on a fixed string on his lyre). Starting from room A_1 at 55 meters, he walked up the stairs until his altimeter read 82.5 meters (55 meters x 1.5—that is, the ratio of 3:2). From this, he deduced that room E on the next floor would be double the altitude at 165 meters, the next double that at 330 meters, and the next double again at 660 meters.

But now he hit a problem. With only an altimeter and knowledge of the 3:2 ratio, how could he plot any other rooms on the first floor? Starting at room E at 82.5 meters, Pythagoras counted on seven rooms and used the ratio again. This was room B_2, which had an altitude of 123.75 meters. By halving that

altitude, he knew where to plot B_1—61.875 meters. He returned to the first room, A_1, walked up the spiral staircase until his altimeter read 61.875 meters, and plotted room B_1. Pythagoras now had positions on his staircase for rooms A, B, and E. Seven rooms up from B_2 was room F♯, and seven from there was C♯ (278.43 meters). As it was his third C♯, he halved it and halved it again— and, hey presto, the genius now had the altitudes of five rooms plotted on his first floor (A, B, C♯, E, F♯). He climbed the stairs to do one more, seven rooms on from C♯, the room G♯$_4$. Running up and down these floors was getting tiring, so he changed tack. He could work out the other six rooms by *dividing* the altitude of A_1 by 1.5 instead of multiplying it. Seven rooms down from A_1 is D at 36.6 meters—he doubled it to 73.2 meters. Pythagoras again walked up the stairs to his first floor and plotted the position on the stairs for room D. Within no time, he knew all the altitudes of the first floor rooms and consequently could plot the altitudes of the rooms on all floors as they rose by a factor of two each time. Put that in your pipe and smoke it, Archimedes!

The Pythagorean Tuning was a roaring success, a mathematical marvel of an infinite cylindrical tower hotel. The staff noticed that rooms A, E, and D were particularly popular (in that order), guests claiming that these had the best views. Nevertheless, Pythagoras's success was celebrated, he won the Ancient Athens Tourist Board "Employee of the Year" award, and the hotel went on to serve guests for hundreds of years. Eventually, though, the Pythagorean Tuning looked very old-fashioned, and a multinational travel company called Equal Temperament bought the property with plans for a major refurbishment.

To give it a more contemporary "vibe," the new designers recalculated the rooms' positions not using the old-fashioned and imprecise 3:2 ratio but rather geometric sequencing. Starting with A_1 at 55 meters, they calculated the adjustments of each consecutive room by multiplying the previous one's height by the twelfth root of two ($^{12}\sqrt{2}$), which is how we now plot our twelve semitones across an octave. The refit proved to be a very costly exercise, as every room except for the As needed to be moved ever so slightly. All room A♯s needed to go up a bit, the B rooms down a little, the Cs up (but not by as much as the A♯s), the C♯s down (by more than the Bs), etc. When the refit was completed, it was undoubted that the rooms were more accurately equidistant; after all, this was what the Equal Temperament company prided itself on. However, guests complained that many of the views were not quite as perfect as they used to be and that the hotel felt a bit banal and corporate. History suggests that equal temperament may not be the last tuning system that Western music adopts, but whatever system we choose, it's always going to be a bit of a botched job.

Tartini's Undertone

Returning to the piano tuner, you might be asking whether a frequency of five beats a second—the by-product of two slightly out of tune strings—could, in theory, be increased to the point where the beats become one continuous audible note. Can two simultaneously sounding frequencies with a pitch difference of 440 Hz, for example, create "beats" capable of producing the note A_4? In 1754, Italian violinist and composer Giuseppe Tartini published some startling findings about sound in his treatise titled "Trattato di musica secondo la vera scienza dell'armonia" ("Second Treatise on the True Science of Harmony"). When he played two notes simultaneously on his violin, he could often hear a *terzo suono*, a third sound. But where was this mysterious extra note coming from? Yes, it was the piano tuners' beats at a frequency high enough to become a single note in the human hearing range. As one travels higher in pitch, difference tones (or Tartini tones) can become more pronounced. This turned out to be rather handy for police work.

The Metropolitan Police Service in London was established by Sir Robert Peel in 1829. "Peelers," as they were known, wore blue tailcoats (to make them look less military than the servicemen dressed in red) and top hats and carried a truncheon, a pair of handcuffs, and a wooden rattle. The rattle was used to attract attention when help was needed or to indicate to local "ruffians" that they should note the presence of a policeman. As the streets of London became busier and noisier, though, the rattle became less effective. During the 1880s, it was replaced by the two-tone whistle. Joseph and James Hudson started their own whistle business in Birmingham in 1870 and were perfectly poised to exploit the police service's increasing desire to be heard above the urban din. It is claimed that Joseph got the idea for the sound of the whistle when he accidentally dropped his violin, as it made a rather disturbing high-pitched discord when it hit the floor. To recreate the sound, he developed a two-note prototype whistle. He submitted it to the Metropolitan Police, who accepted it as the best of the whistles that had been offered to them. Indeed, the present-day J. Hudson & Co. (now called ACME Whistles, whose products are used by all kinds of sports referees worldwide) states that "when first tested on Hampstead Heath in London, it was heard over two miles away." The pitch of the whistle is in a frequency range much higher than that of most Victorian urban noise. But one cannot help wondering whether the Hudson brothers were aware of quite how distinctive and startling "the Metropolitan" whistle really was. This is partly due to the difference, or Tartini tone, that was accidentally built into the design. Joseph combined two pure high-pitched notes relatively close together in frequency, creating a sound shrill enough to gain everyone's attention.

I bought my own replica Metropolitan whistle and, much to Cysgu the cat's disgust, experimented with it at home in my studio. After recording a number of very hard whistle blows and analyzing the resultant frequencies in my recording software, I noted the following: first, the cat left the studio; second, the two whistle tones are approximately A_6 at 1,820 Hz (two octaves above an oboe's tuning A) and C_7 at 2,150 Hz; third, my ears really objected to the distorted *terzo suono* difference tone of 330 Hz, which I recognized as an E_4 (officially 329.63 Hz), the first note Chubby Checker sings in his 1960 hit "The Twist." The Hudsons had truly hit on an incredibly distinctive and unnerving sound, not created by just one shrill interval but a combination of *two* intervals, the second created by Tartini's mysterious difference tone.

Once the whistle was approved by the Metropolitan Police, initial orders around 1884 totaled twenty-one thousand in several batches, and J. Hudson & Co. went into overdrive. The Metropolitan two-tone whistle (should it not now be called the three-tone?) soon became a ubiquitous piece of equipment, adopted by other services, including the fire brigade and the army. Years later, it provided the accompaniment to "going over the top" during the trench offenses of such notable battles as the Somme in the First World War. The infantry training manual of 1914 informs troops of the whistle's "Cautionary Blast (a short blast)," "The Rally Blast (a succession of short blasts)," and "The Alarm Blast (a succession of alternate long and short blasts)." What the authors of the manual could not have imagined was the gut-wrenching sound of hundreds if not thousands of these

orchestrated blasts across miles of Allied trenches—a cacopho-
nous, fatalistic overture for countless soldiers who climbed out of
the mud to certain death.

The opposite of Tartini's undertones are overtones. These
are present in all musical notes we hear (and even the ones we
can't). As discussed in chapter C, overtones are like the individual ingredients of a beef Bolognese recipe. The tone of a trumpet
note is defined by its unique blend of overtones, different from
the one that produces the same note on a guitar. But what the
instruments share in common is that the overtones follow the
same pattern in nature: in other words, the ingredients are added
in the same order every time, regardless of their quantities. Using
the note A_2 at 110 Hz as a fundamental (the low guitar note that
opens Norman Greenbaum's "Spirit in the Sky"), its overtones
are found at the following ratios:

1st overtone, 2:1 ratio, A_3 (220 Hz)
2nd overtone, 3:1 ratio, E_4 (330 Hz)
3rd overtone, 4:1 A_4 (440 Hz)
4th overtone, 5:1 $C\sharp_5$ (550 Hz)
5th overtone, 6:1 E_5 (660 Hz)
6th overtone, 7:1 G_5 (770 Hz), and so on.

All these notes are present in a single musical note played
by a clarinet, tuba, or guitar, but different levels of each overtone allow us to recognize the sound of that particular instrument. Although these overtones are a handy "fingerprint" guide
to instrument recognition, they also bring inherent problems for
composers. Though the likes of Vivaldi, Bach, and Chopin probably didn't understand the physics behind overtones, they were
acutely aware of their pitfalls: combine two low pitches, and the

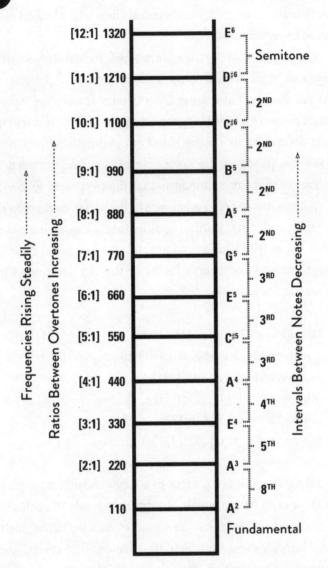

The Overtones of an A^2

audible overtones quickly pile up. If Vivaldi were to write a low A for double bass at 55 Hz and a C♯ for cello at 69.3 Hz, theoretically they should blend very nicely, as the interval distance between these two notes is a major third, a happy-sounding consonance. However, their blend is compromised by an almighty car crash of overtone frequencies (55 Hz, 69.3 Hz, 110 Hz, 138.6 Hz, 165 Hz, 207.9 Hz, 220 Hz, 275 Hz, 277.2 Hz, 330 Hz, 346.5 Hz, 385 Hz, 415.8 Hz, 440 Hz, 485.1 Hz, 495 Hz, 550 Hz, 554.4 Hz, 605 Hz, 623.7 Hz, and so on). And that's before adding the final note of this "happy" major triad, perhaps a low E on viola. The resulting chord, far from being a summery A major chord of pure sunlight, would be a grumbling, muddy mess with far too many overtones competing for our attention. With such a chord, Vivaldi would not create a perfect Bolognese recipe but an overly rich mishmash where too many ingredients compete rather than complement. This is why bass instruments such as the double bass and cello tend to play the same notes an octave apart in baroque and classical music. The mixing of notes in chords tends to be placed higher up in the frequency range, where many of the problematic overtones that instruments produce disappear above our audible limit. Vivaldi, along with most composers, simply went along with what sounded acoustically nice, even when portraying the depths of winter.

To Each Their Own

Moving the notes of a chord around to produce a pleasing mix is called "voicing." All composers with any craft "voice" chords. I am confident that Frédéric Chopin, the nineteenth-century Polish composer and piano virtuoso, did not initially list the gamut of potential overtone frequencies before embarking on writing the opening low-pitched left-hand chord of his Nocturne in E Minor (E at 82.4 Hz, G at 98 Hz, and B at 123.5 Hz). He knew that this low chord sounds like a messy grumble. His solution was intuitively clever and relatively simple—to move the middle G (the note that shares the least number of overtones with the other two) one octave higher, thereby reducing the number of that pitch's overtones audible to us and avoiding many overtone clashes.

Over the centuries, composers, orchestrators, and arrangers have been the craftsmen and women who have creatively and ingeniously steered musical ideas through an incredible obstacle course of frequency pitfalls, like skilled bumper car drivers avoiding overtone head-on crashes. With this in mind, all instruments have specific and important roles: violas in orchestral music and acoustic guitars in rock and pop provide the custard in the musical trifle; flutes, oboes, and synth leads are the creamy bit with sprinkles; and double basses and cellos should be treated with care down in the spongy Jell-O bass range. This is precisely why musicians with delusions of grandeur—such as virtuosic bass guitarists—are wasting their lives trying to recast their musical roles so that they may appear in the limelight. Know your role, bassists—stay out of the way, and if you want to show off and be the center of attention, go and learn one of the creamy instruments, the decorative ones that occupy the higher frequency range.

In terms of studio recording and mixing, anything to do with frequency is addressed through EQ, or equalization. One of the most challenging parts of studio EQ is the treatment of the recorded voice. Of all musical instruments, this is undoubtedly the most difficult to capture and equalize successfully as it is constantly changing. Whereas a violin has a relatively stable timbre (its frequency and overtone "fingerprint"), the addition of phonetics on top of the characteristics of an individual's voice can make EQ-ing vocal recordings an absolute minefield.

When less experienced singers first enter my studio, they're always interested in the pop shield or filter, that trendy circular piece of mesh or gauze placed in front of the microphone between the singer and said mic.

"It's a pop shield," I tell them, bracing myself for the same sort of response each time.

"Cool. It looks really cool when you're singing pop," or "But I want to sing crust punk, not pop."

It's important to make a singer feel comfortable in what can be a very daunting and stressful environment, so I suppress my world-weary expression, grin inanely, and launch into a mini lecture on the science of phonetics. So here goes. I'm grinning inanely while I write.

A pop shield has nothing to do with pop music. The shield is usually made of layers of stretched nylon material that mitigate the effects of the way we produce the sounds of certain letters of the alphabet, collectively known as plosives. If you have ever listened to a speech poorly delivered through a microphone, you will instantly recognize a plosive. The nervous orator often fails to understand that the microphone is there to amplify their voice to the far corners of the venue. For some, it becomes a life raft, something to cling on to, something to stoop right up to and deliver their piece at point-blank range. For those on the receiving end, the sporadic low frequency booms coming from the PA system make for a very arduous experience. The booms are the result of the bursts of high-speed air we release from our mouths when creating b's and p's (bilabial plosives), t's and d's (dental plosives), and k's and g's (velar plosives). The bilabial plosives are the main culprits, as the delicate diaphragm in the microphone is shocked by the sudden air burst—this is called "popping." The pop shield attenuates or weakens the energy of the air burst, reducing the impact on the mic's diaphragm.

Such technical EQ challenges often inspire new, innovative means of artistic expression. As well as plosives, there are other problematic vocal sounds. The letter *s* is a sibilant, a type of fricative consonant. Its nearest relative is the letter *z*, both produced by directing a stream of air between the teeth. Sibilants have a higher frequency range—when saying the word *hertz*, the combined *tz* sound has predominant frequencies between 4 kHz and 8.3 kHz. This range of the audio frequency spectrum is where the sparkle begins, equivalent to the touches of white paint on the crests of an artist's aquamarine waves in a seascape. Music without these higher frequencies sounds like it is underwater or in the next room. To experience how high frequency has an impact on the excitement and dynamism of music, listen to the opening few seconds of Modjo's 2000 hit "Lady (Hear Me Tonight)." Within four bars, the track sweeps from dull to sparkling light, simply through the addition of higher frequencies (or "treble," as it was called on your average old-school stereo system). One of the pitfalls of too much treble, though, is that the music can start to sound thin and brittle. Sibilants exaggerate frequencies in this range and can be a major cause of discomfort and listener fatigue, which is why most decent audio recording software comes with a "de-esser" plug-in (this small app cleverly "softens" the high frequencies of an *s* sound). But the creative songwriter, singer, and producer George Michael turned this hissing sibilance into a calling card for his music. Albums such as *Older* are awash with vocals drenched in long reverberation, making a creative feature of George's *s* and *z* sibilants. Through a system of noise gates (which only let certain frequencies and volumes be heard) and reverbs, George Michael and his engineers cleverly refashioned a potentially annoying vocal problem into a cool, inventive, and unique sound that is now often dubbed the "George Michael reverb."

Shutting Out the Noise

Our story of the combination of frequencies has until this point been one of addition—whether that be the addition of beats, undertones, or overtones. But on occasion, mixing waves together results in a reduction, a basic physics principle that some physicists and entrepreneurs have exploited to make our traveling lives significantly more comfortable and pleasurable. Combining two waves of the same frequency at the same time results in the point of highest air pressure (compression) being increased and the point of lowest pressure (rarefaction) being decreased. But start one of the waves slightly later than the other, and the point of compression of the first wave can be canceled out if it occurs at the same moment as the second wave's rarefaction—while one pushes, the other one pulls. Similarly, when the first wave hits its point of rarefaction, the second wave will be at the point of greatest compression, and again, it will cancel out the effect of the first wave. In other words, one can fight sound with sound. And this is the principle on which noise-canceling headphones work. The story of this technological marvel is populated by great theorists and inventors, many of whom developed the necessary science, some of whom are claimed to have been the main mind behind the principle. From Paul Lueg in Germany in the 1930s to William Meeker and the brilliant polymath Dr. Lawrence Fogel in the United States in the

1950s, they have explored the ability to silence sound with more sound, a theoretical and practical reality for decades—reserved, though, for an elite few, such as military pilots. The incident that sealed the commercial potential of noise-canceling headphones was a very, very long flight in 1986, from the Mojave Desert all the way to...the Mojave Desert, nonstop around the world in under ten days.

Eight years previously, on a considerably shorter commercial flight between Switzerland and the United States, MIT professor, acoustic engineer, and businessman Dr. Amar Bose was given a pair of headphones, an exciting innovation in in-flight entertainment at the time. Dr. Bose did not particularly enjoy the music on offer, as it was spoiled by the considerable noise of the aircraft cabin. Apparently, by the end of the flight, Bose had sketched out his plan for a pair of active noise-cancellation headphones. And on December 14, 1986, wearing Bose's prototype, pilots Dick Rutan and Jeana Yeager sat in their cramped Rutan Model 76 Voyager aircraft as it trundled down the five-kilometer runway at Edwards Air Force Base at the start of their attempt to fly nonstop around the world. One could argue that Rutan and Yeager were perhaps being a little precious by insisting on such a luxury. However, as the aircraft needed to be as light as possible, it had next to no sound insulation, and both pilots were never more than inches away from the engines. The noise level in the cabin was recorded at 105 decibels (dB)—that's like someone banging a large drum next to you, for ten days. The active noise-canceling part of the headphones failed around halfway through the flight, which left the pilots with only passive noise cancellation (that's just the foamy muffly stuff). The *Los Angeles Times* reported that the pilots should have been experiencing between 80 and 85 dB if their headphones had been working and that the ground crew were becoming increasingly worried about fatigue—a serious effect from prolonged exposure to loud noise. The pilots made it in the end, setting a new world record that remains unbroken. Through their very uncomfortable and grueling endurance flight, we can now—thanks to the work of Dr. Amar Bose and Dr. Henning von Gierke, along with engineers at the Sennheiser company in Germany—effortlessly cancel out all forms of exterior noise.

The principle of active noise cancellation is relatively simple, the practicalities a whole lot more complex. A microphone on the outside of the headphones records extraneous frequencies, perhaps the persistent low rumble heard in an aircraft cabin, between roughly 40 Hz and 100 Hz (the same range as a bass drum). It registers an incoming sound wave (for example, 40 Hz), then plays that same wave through the headphones to the listener. However, it inverts the wave, placing it out of phase so that the compressions and rarefactions are exactly opposite, canceling out the sound of 40 Hz coming from the aircraft cabin—the rumble miraculously disappears. Active noise-canceling headphones work across the whole audible spectrum with an astonishing response speed, leaving us to enjoy an undisturbed listening experience.

Disco Inferno

This form of destructive combining of waves is equivalent to fighting fire with fire. And two former students of George Mason University in Virginia have taken the concept of destructive waves into just that area. But they do not fight fire with fire—they fight it with sound. Seth Robertson and Viet Tran's final project of their engineering degrees was a fire extinguisher that uses no water or toxic chemicals, just sound waves. As you might imagine, it was met with a certain level of skepticism among their fellow students and many academic staff. Fortunately, some influential people stuck by them, and following their successful graduation, they set up their own company, which continues to develop sonic fire extinguishers. Many before Robertson and Tran have tried this (including the military), but they tended to focus on ultrasonic waves. Robertson and Tran found success using lower frequencies, which have larger wavelengths and an ability to separate the oxygen from the fuel. From the online demonstrations, the extinguisher seems to be blasting out a frequency around 39 Hz, the E♭$_1$ that dominates Stevie Wonder's bass line in "Superstition" (it's played on a keyboard in the original—the note is too low for a standard bass guitar to play). The implications for Robertson and Tran's extinguisher are considerable—conventional extinguishers cause water and chemical damage to areas around a fire and are very problematic in the microgravity of a spacecraft, for example. There are still issues to be solved, in that the heat of the fuel source remains after the sonic extinguisher is switched off, thereby increasing the chances of the fire reigniting. However, the idea that genius bassist Esperanza Spalding could put out a small studio fire by fiercely plucking her detuned E string is an intriguing one.

Back in the world of combining frequencies in music, that bass range of the sonic fire extinguisher can be considered problematic when mixing. For many years, I have taught my students to filter out frequencies from 20 Hz to around the 30–40 Hz range. Following a recent conversation with leading engineer, mixer, and producer Dom Monks (his CV includes working with Sir Tom Jones, Coldplay, Sir Paul McCartney, and Nick Cave), I reflected that I rarely ask the question "*Why* do I teach this?" I always assume that others (especially those with more confident attitudes) know more technical stuff, and I willingly defer to some unknown, unwritten studio rulebook that decrees that practices such as cleaning up fluffy and rumbly unwanted frequencies are a must. But Monks questioned my assumptions, my sense of technical inadequacy, and my often lazy approach to teaching (which manifests itself as the transference of pseudo rules without interrogation of their legitimacy). Combining frequencies in a studio setting through mixing and mastering is often seen as a science. The internet is bursting with producers, mixing and mastering engineers, as well as audio companies feeding on the paranoia of studio musicians such as myself, offering rules, tips, and expensive plug-ins. Dom Monks is skeptical of much of it. "Companies say, 'Here's a conga EQ preset, press this setting for your congas.' Hang on, it doesn't know what they sounded like. The preset has no idea how you recorded the congas—it makes no sense at all," said Monks. "People often say to me, 'Ooh, I don't know how it sounds technically,' to which I'm always saying, 'It doesn't sound anything *technically*.' Through my theoretical training, I can tell if there's an excess of a certain frequency, but what is an *excess*? That's just my opinion about that relationship. To say to someone, 'You've got too much of this frequency here,' well, maybe for one person, but for another, that's exactly how they meant it to be."

I asked him whether you must not have any frequencies from around 30 Hz downward in your mix. He responded, "You can literally do anything. It's like fashion—someone says, 'You can't wear that with that,' then someone wears that with that, and that's the thing!"

Monks cited the Billie Eilish album *When We All Fall Asleep, Where Do We Go?* as an example. "People say, 'There's no top end in this at all!'...and there isn't. Look at the 101 rules for pop music—top end is a must, apparently, but here's a big hit!" He concluded, "Your only real reaction is an emotional one of 'Does this move me, or does this not?' If not, why not?" And Monks is right. The mind-boggling combination of a vast array of frequencies, fundamentals, overtones, undertones, and beats that make up even a single chord is for nothing if it fails to elicit an emotional reaction. After all, that is the main purpose of music.

ONLY THE LONELY

Vibrations in the
natural world

At the height of the Cold War, the Americans had their ears to the ground all over the globe. One particular area where the Soviet threat loomed large was the deep ocean. Russian submarines such as the Whiskey class could roam the seas undetected, sneaking up close to the West's shipping and potentially even taking a periscopic peek at bathers on a Florida beach. When Soviet submarines started to carry nuclear-tipped missiles, the need to track their every move became of existential importance.

Throughout the 1950s and 1960s, the U.S. Navy invested a great deal of time and money in the highly classified SOSUS (Sound Surveillance System). They dropped long lines of ocean microphones—known as hydrophones—into both the Atlantic and Pacific and listened for the low rumbling engine noise of Soviet submarines. In the 1940s, scientists had discovered that parts of the ocean are astoundingly good at transmitting sound over hundreds and even thousands of miles, especially at low frequencies. You could, in theory at least, play a tuba in the waters around Cornwall and hear it off the coast of Newfoundland. The farther away the Americans could hear enemy submarines, the more time they had to prepare for any Soviet shenanigans.

To understand this phenomenon, picture a cross-section of the ocean as a slice of layer cake. The layers, or channels of water, are created through different combinations of pressure, temperature, and salinity. One such layer around one thousand meters deep is named the deep sound channel or SOFAR (sound fixing and ranging). This layer traps low-frequency sound, bouncing the waves off the roof and floor of the channel back toward the middle, thereby creating a long-distance acoustic effect. It was here that the U.S. Navy submerged long lines of hydrophones as part of their antisubmarine warfare strategy. But of course they didn't just hear ominous rumbles from the likes of Red October. They heard all the sounds of the deep ocean, beautiful and mysterious acoustic wonders that have been the soundtrack of the seas for millions of years. A whole new sonic world was revealed, emanating from sources as varied as earthquakes, volcanoes, rain, and bubbles as well as marine life like dolphins, oyster toadfish, and snapping shrimp.

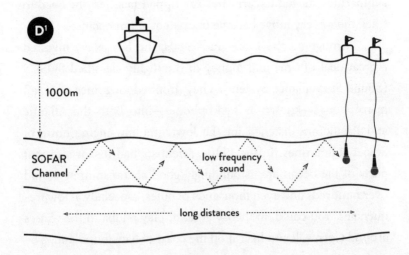

One man was especially important in bringing the sounds of deep ocean life to the world's ears. William Watkins joined Woods Hole Oceanographic Institution in Massachusetts in 1958, and although he was not the first to record underwater sound, his electronics expertise meant he could build robust portable tape recorders that could withstand the rigors of ocean travel. Watkins spent four decades at Woods Hole recording a vast array of aquatic mammal sounds. Of these, the voice of one type of mammal came to enchant the whole world with its haunting and plaintive cries—the whale. Watkins recorded numerous species of whale, each of which have their own distinctive songs. However, one whale call proved to be even more than species unique. This creature was completely unique. It became known as the 52-hertz whale (or 52 for short).

Hello?

In 1989, SOSUS started recording a distinctive whale call whose frequency was measured between 50 and 52 Hz. The lowest G♯ on a standard orchestral double bass is 51.9 Hz, one small musical step flatter than the opening note of the famous bass riff in Fleetwood Mac's "The Chain." It's about seven white notes up from the lowest note of a standard piano. This is quite a surprise, as the whale songs most of us would recognize inhabit a much higher frequency range, giving them their haunting, plaintive, eerie, almost child-like quality, but this is because whale song is often pitch-shifted after it has been recorded, transposed up into the frequency range occupied by music.

How did Watkins and his colleagues know this was a whale call? They had never seen the mammal; they had only its call to go on. A low rumble of a bass G♯ could have been anything in the ocean—perhaps a Soviet submarine, or even a tuba player in Cornwall. And how did they identify it as a biological sound, let alone a *whale* call? Watkins, Mary Ann Daher, Joseph George, and David Rodriguez published their findings in 2004 after recording 52's calls for fifteen years. There were six defining attributes of the calls that led them to conclude that this was indeed a solitary and unique whale: the calls were all very good recordings collected from many hydrophones over many occasions; the frequency range across the whole fifteen years was very small, between 50 and 52 Hz; the calls were consistently between three and ten seconds long; there was a consistent pitch drop at the end of the call, by up to 2 Hz; they were repetitive grouped calls; and finally, the calls were always only from one source, never with any overlap. It couldn't be anything but a whale.

When it was announced, the media had a field day, calling the mammal "the loneliest whale in the world"; the public took the creature to its heart. Imagine roaming the vast expanse of central and eastern parts of the north Pacific basin, calling out to other whales, advertising for dates with the promise of candlelit meals, cozy nights in, intimacy, and sunset walks/swims, etc., only to find that no one ever responds. It's not that other whales couldn't hear him or her; they could. They just didn't understand the language he or she was speaking. Blue whales and fin whales call between roughly 10 and 25 Hz, seventeen notes down into Infinite Piano range; 52 Hz, that low, out-of-tune G♯, is just whale gobbledygook, cetacean double Dutch.

We can all anthropomorphize an intelligent, beautiful, and majestic mammal desperately swimming the endless ocean, calling out for fifteen years, unaware that its peers could not understand it. And this whale most definitely traveled far and wide. Its route over the years seems to suggest the whale's increasing desperation. During the first three years of recordings, 52 remained around one location, 46°N, 126°W. But then it started wandering in what seemed like a more and more hopeless search for company; in 2003 and 2004, it traveled over eleven thousand kilometers. Watkins's paper reported that there were no "apparent repeated patterns" to the whale's travels and that "no relation to any other whale movements could be traced."

The uniqueness of the calls of 52 provided a unique opportunity for the scientists at Woods Hole Oceanographic Institution. The problem with trying to keep tabs on a "normal" whale is that sometimes it goes quiet. When its calls restart, there is no way of proving that it is the same individual. But with no other precise 52 Hz call anywhere in the ocean, following this lonely individual was possible.

There is some mystery about what species of whale 52 is, because it has never actually been spotted. Experts have suggested possible answers but nothing definitive. Bizarrely, it could be a hybrid, as whales do seem to be attracted to other species. A female fin and blue whale hybrid was caught in 1984 off the northwest coast of Spain. She had a blue mother and a fin father, and her size was in between the usual parameters for the two species. And in 2011, even more remarkably, a hybrid minke whale was discovered. DNA testing on this Norwegian-caught northern minke revealed its mother was an *Antarctic* minke. These whales should have been hemispheres apart, but there it was, a crossbreed between northern and Antarctic whales. Alternatively, the lonely whale could be an anomaly, a "freak of nature." Statistically, it's difficult to believe that across the globe's vast oceans, there could be anything biological that was completely unique, but Watkins seemed convinced.

As the years have gone by, 52's calls have become more

mature. Just like human singers, the high notes get more difficult with age. The whale's calls have descended, only by about 2 Hz, but perceptible enough to be of note. Just like Maria Callas or David Bowie, though, the year-on-year struggle for those youthful high notes hasn't made the performances any less attractive; Maria Callas would transpose arias and songs down by a few semitones on her tours in the early 1970s, and in his later years, Bowie's live versions of "Life on Mars" went down by as much as five semitones.

The Rowdiest Race

It's not only that 52 has struggled vocally though; there has been a growing struggle for *all* whales to be heard in an increasingly noisy ocean. It doesn't take much of a guess to deduce who is responsible for this crescendo of oceanic background noise—us humans. The glamorous term is *anthropogenic noise*, but in reality, it's just another depressing by-product of what we call "progress." Whether it is the sound of ships' engines, sonar, industrial maritime exploration, or even underwater explosions, we are tramping our way through the oceans just like we've tramped our way over the land, inflicting substantial damage on those who have inhabited their environments for millennia.

Research in 2012 by the Scripps Institution of Oceanography in San Diego revealed that baleen whales reduced their calls by half when faced with sonar between 1,000 Hz and 8,000 Hz. The former is around the pitch C_6, the heavenly high note sung by a solo treble voice in Gregorio Allegri's *Miserere*. To play a note at 8,000 Hz on our Infinite Piano, one would need to travel eight white notes (an octave) beyond the upper limit of a standard piano. But why would baleen whales, whose calls are all below 100 Hz, be affected by sonar ten times higher in the frequency spectrum? Answers are not clear-cut, but the researchers at Scripps deduced that, at the very least, whales can hear in this range. Why would they need to hear up there when all their friends call at the same low frequencies as they do? Some have suggested that perhaps the answer lies not in who they wish to attract but who they want to avoid. Killer whales produce vocalizations in parts of this frequency range. They often prey on blue whales, so to avoid being on the evening's menu, an ability to hear an approaching attacker is essential.

It was recently estimated that, with the advent of modern shipping, ocean noise in the low frequency range (30–100 Hz) has doubled in each of the past four decades, adding to the ever-increasing "acoustic smog." Christopher Clark, who ran the Bioacoustics Research Program at Cornell University, likened whales' current acoustic environment to living on the tarmac of a busy international airport. Low-frequency noise from ships is drowning out whale calls, including those of 52. Another notable oceanic noise pollutant is low-frequency active sonar, which is used primarily for military purposes. Because it is often blasted at high volumes from lines of up to eighteen loudspeakers hundreds of meters down in the ocean, it travels immense distances. U.S. Navy tests have made signals clearly audible across the entire north Pacific. Such sonar has profoundly disturbing effects on ocean creatures, particularly mammals.

Whales, like many animals competing with human noise, have been found to use the Lombard effect. This is the involuntary action of modifying one's voice when communicating in a noisy environment. Humans unconsciously speak louder where noise has the potential to mask their ability to be understood—think cocktail parties, sports stadiums, restaurants, and nightclubs.

And it's not only the volume that is modified by this involuntary action. We also change the speed of our words, the duration of syllables, and the frequency at which we talk. Humans and whales are not alone in shouting louder and pitching higher to get heard. Studies of blackbirds have shown that city birds use the Lombard effect, singing at a higher volume and frequency than their rural counterparts. One such experiment compared blackbirds in the woods around Vienna with those that reside in the city. The findings revealed that the most common peak frequency for the forest sample was between 1.8 and 1.9 kHz (1.8 kHz is the high fifth note of the opening phrase of "Dawn" from Benjamin Britten's *Four Sea Interludes* from Peter Grimes), whereas the city birds' peak frequencies were 2.3–2.4 kHz. This difference between the rural and urban birds' pitch is about a perfect fourth interval, the distance between the first two notes of Mendelssohn's "Wedding March." Singing at a higher pitch allows the birds to increase their volume and also raise their songs further away from the predominantly low-frequency traffic noise of the city. It is believed that this helps blackbirds to reduce the impact of "anthropogenic acoustic masking"—in other words, human-made noise.

Making a Silk Web out of a Spider's Ear

Anthropogenic noise doesn't only affect those who call and sing. Even the humble garden spider has been impacted by our inability to do anything without making a big racket. A 2014 study suggests that human-made noise significantly impacts the ability of the web-building European garden spider, *Araneus diadematus*, to detect prey. Not only does background noise confuse spiders' acute vibrational sensitivity, but the study also found that human-made substrate (substrate is anything a creature stands on, perhaps a leaf or soil, but in this case, tarmac, concrete, etc.) reduces the volume of the vibrational signals that the spiders receive. Human activity is having a catastrophically detrimental effect on spiders' feeding habits.

As spiders have, on average, eight eyes and no sticky-out ears, it has always been a fair assumption that vision has been higher up their list of sensory priorities than hearing. Recently, though, it has been discovered that spiders "hear" through unlikely parts of their bodies, and it turns out that frequency has a very important part to play in their survival. Anthropogenic noise is as detrimental to spiders as it is to whales.

In 2016, Cornell University published startling arachnid-related news. Their subject was a species of jumping spider called *Phidippus audax*. Though the research paper's title, "Airborne Acoustic Perception by a Jumping Spider," is perhaps a little dry, the journey to its findings reads like a gripping scientific "Eureka!" story packed with mystery, abject failure, accidental discoveries, and gruesome death.

Jumping spiders have always been thought of as primarily visual creatures as they don't have tympanic ears (eardrums). They can, though, sense air movement through hairs on their legs. This is called "near-field" hearing, where moving particles stimulate the hairs by tiny distances, in the micron range. But how does one test whether a spider can actually hear something? Observing its behavioral response to a sound is one plausible test, though it's not particularly conclusive, as the spider might not be *hearing* anything. The other way is to get into the brain of a spider such as *Phidippus audax* and observe any neural reactions to sound. There are two major drawbacks to "getting inside" such a brain: one, it is the size of a poppy seed, and two, the spiders' bodies are pressurized. The moment one makes an incision, the spider's body has a "blow-out," and it tends not to hear anything ever again. This proved to be most unhelpful to the researchers.

But one scientist, Gil Menda, found a way of holding a jumping spider in a little box and inserting a self-sealing tungsten microelectrode (a tiny needle) into its brain without the creature going "pop." As Menda told me, "The spider was almost never damaged, and I could actually take the electrode out, seal it, and the spider could live a full life after that." At first, he noticed the visual system of spiders—"the neurons are talking to you; they fire electrically with different frequencies. I could hear their brains firing in response to visual stimulus… When we put the electrode deeper in the brain, we got to an area which was responsive to sound." Menda and other members of his team then played frequency sweeps between 50 and 400 Hz through a loudspeaker, monitoring how the neurons of the spider's brain fired as it "heard." Frequencies around 80 Hz seemed to trigger the most response—this is the lowest E string on a guitar. (Have a listen to the first note of the famous classical guitar arrangement of "Malagueña" composed by Cuban Ernesto Lecuona, or the opening of Dick Dale's 1960s surf classic "Misirlou," which was used as the theme for the film *Pulp Fiction*. The E at the start of both of these is at 82.4 Hz.)

What happened next in the spider laboratory surprised everyone. Menda related the story to me: "I moved my chair backward, and it made a noise in the room. I heard suddenly the spider's brain respond very loudly to my noise. Then I tried to do more noises." Menda started to clap at the spider, not in appreciation of its talent but because a clap is a multifrequency sound. Amazingly, the neurons fired again and continued to respond up to a distance of almost five meters away. "I wasn't the spider expert. I was the neurophysiology expert," said Menda, "so I called to my buddy and asked him for information about what a spider can hear. 'No, they cannot hear,' was his response. I asked him to clap his hands, and the spider's brain made a noise. All the people in the lab came. It was a moment of eureka!" This was now a groundbreaking biological study. They transferred the spider and the experiment to an anechoic chamber (a room purposely designed to have no extraneous noise or reflections in it) and produced similar results. Jumping spiders' brains responded to airborne sound between 80 and 380 Hz (a frequency just above the top E string of a guitar), in both far-field and near-field ranges. In other words, jumping spiders can hear anything in a guitar range between the lowest and highest strings. Perhaps I might suggest a new name for it: *Phidippus audax telecaster*?

There was a further part of the experiment that was even more revealing. As well as studying the neurophysiological part of spiders' hearing (the brain stuff), the team wanted to explore further, focusing on spider behavior. A wire mesh cage on top of an extremely heavy metal block—through which most vibrations could not be passed from the ground—was constructed. A single spider was placed inside the cage. When a tone of 80 Hz was played, the spider almost instantaneously froze—called an "acoustic startle response." (Try it at home. See if your *Phidippus audax* stops dead when you play it the first guitar note of the "Peter Gunn Theme.") When a frequency of 2,000 Hz was sounded, there was no reaction; the spider simply carried on in its own spidery way. Whenever they sounded the 80 Hz though, *Phidippus audax* froze. This was perplexing, until a member of the team made the connection with dangerous predatory flying insects. Some species of wasps, during flight, produce dominant acoustic frequencies in a similar range. Possessing the ability to hear a predator gives the spider an early-warning survival tool, just as a blue whale needs to hear an approaching killer whale. Sensitivity in this range also gives the arachnid (or "lunch," as it is known to certain wasps) the ability to hear the movement of leaves, snapping twigs, or other aural signals of larger predators. As well as spider defense, hearing in this range gives it the upper hand when trying to catch its own food; many flying insects that the spider is partial to have wing beat frequencies across this spectrum—for instance, the fruit fly at around 200 Hz.

The team also investigated the hairs on the spiders' legs. They performed "direct mechanical stimulation" of hairs on the patella of the foreleg. In other words, they tickled the spiders with the fabulously named "linearly actuated micro-shaker," a very mini tickling stick. Once again, the neurons in the brain responded to low-ish frequencies between 64 and 256 Hz. Across all these experiments, scientists learned the following: spiders can hear airborne sounds; they hear through their legs; they are particularly sensitive to frequencies that their predators generate; and perhaps most surprisingly, they can hear over long distances...and, oh yes, they have ticklish legs.

But this is far from the end of the story of spiders and frequency. Just like humans, many spiders dance as part of a courtship ritual. However, the similarity probably stops there, as in the arachnid world, there is a high risk that unsuccessful males will be murdered, liquefied into a soup, and consumed. Male peacock jumping spiders have the most elaborate of courtly dances to attract females. It's not just leg-waving moves and brightly colored fan displays that are used in their wooing boogies. The ladies are often initially alerted to a pursuing male via substrate-borne vibration. Male peacock spiders tap their feet on substrate, creating rapid vibrations as soon as they sense a female. They have a wonderful dating repertoire including *rumble-rumps*, *pre-mount crunch-rolls*, and *grind revs*. Scientists are not yet sure how all these are produced, but getting the mood right—once again, around 80 Hz seems to do it—can mean the difference between life and death, depending on whether the female senses bad vibes or good vibrations. Through the power of dance and drumming (which involves waving their legs in the air, stomping back and forth to the right and left, shaking their abdominal iridescent fans, and wobbling their tummies), male spiders are very keen to strut their Casanovan credentials.

Research on webs has also produced startling results regarding spiders' reliance on vibration. Between 2014 and 2016, the delicately named Oxford Silk Group—part of the Department of Zoology at Oxford University—carried out extensive research on the silken properties of webs. Web-based spiders rely on the multifrequency vibrations that struggling flies and other bugs transmit along the silky strands. Strings of silk can be tuned by spiders, just like guitar or violin strings. Beth Mortimer, one of the Oxford Silk Group, said, "The sound of silk can tell them what type of meal is entangled in their net." As well as receiving vibrations, the team discovered that spiders "pluck" their silk strings to ascertain the condition of their delicate death traps.

For us humans, spiders' musical connections go back centuries and once even inspired their own dance craze. Asked to name a spider, most people would probably opt for "tarantula." The name originates from Taranto in Italy, where the local spider (actually a wolf spider) was named after the province. Other tarantula-inspired words then followed, including the strange mass psychological phenomenon called tarantism and the folk dance the tarantella. The dance is a frenetic affair with a dum-dee dum-dee dum-dee dum type of rhythm. Most musicologists explain it as a dance in 6/8 time, but this makes little sense to nonmusicians, so best to think in terms of dums and dees. Legend has it that during medieval times, a tarantula's bite would send a Tarantino (yes, that's what they call a resident of Taranto, and yes, it could be a scene from one of his films) into a frenzied state of tarantism, the only antidote being to dance it out of one's body. Onlookers who hadn't been anywhere near a spider would soon convince themselves that they had the spiders' venom inside them too, leading them to join the dancing throng. Instead of witnesses at the scene shouting "Get a doctor!" victims might have heard the words "Get a guitar!"—which was probably less reassuring. This strange cure to the venomous bite caught on for a while, and though the popularity of the antidote soon waned, the dance remained. Nowadays, not much in the way of collective hysteria occurs even during the most notable of all tarantellas, the fourth movement of Franz Schubert's quartet *Death and the Maiden*. It has those most typical tarantella dum-dee dum-dee dum-dee dum rhythms and a freneticism worthy of any death dance. Schubert knew he was dying during its composition, and there is more than a whiff of a ranting outburst at the grim reaper in this final movement.

The grim reaper has long since dispensed with Schubert and now seems to have turned his focus on the entire spider population. A joint 2014 study by scientists from the UK, United States, Brazil, and Mexico suggests that while the human population has doubled in the last thirty-five years, the population of invertebrates has fallen by around 45 percent due to something called "anthropocene defaunation," a mixture of habitat loss and climate change. For arachnophobes and creepy-crawly haters, less bugs in the world may seem like no bad thing. But estimates suggest the world's population of spiders eats at least four hundred million tons of food each year; that's equal to the amount of meat and fish the entire human population gets through per annum. We need as many spiders chomping through as many bugs as possible; otherwise, we could all suffer in the end.

The Waggle

Another little creature on which we rely heavily is also in catastrophic decline—the bee. Over recent years, bee populations have been threatened by a variety of human activities, and anthropogenic frequencies have added to their problems. Distribution of electric power uses alternating current (AC), which in many countries of the world has a frequency of 50 Hz, meaning that the direction of flow of electrons in the current alternates fifty times per second. Overhead power lines distribute this electrical energy over large distances, generating a type of pollution called extremely low frequency electromagnetic fields (ELF EMF). As bees spend a lot of their time on foraging flights, they are often exposed to relatively high levels of ELF EMF around power lines. A 2018 study published in *Scientific Reports* suggests that this form of wave pollution has "the potential to impact on [bees'] cognitive and motor abilities." And when they return from their foraging (whether near power lines or not), they must communicate their findings with others in the hive—cue "the waggle dance."

The dance is used to communicate to other honeybees the whereabouts of food sources outside the hive. Honeybees walk in a figure eight, buzzing their wings and waggling their abdomens. In the bee world, this is the equivalent of GPS instructions, such as "make a left turn at the slightly bedraggled rosemary bush owned by the neighbors who have an old sofa in the garden, fly for another thirty feet, and the nectar is in a bunch of foxgloves to the right of the gas station." For many years, biologists—especially apiologists—assumed that bees, just like spiders, hear very little to absolutely nothing. But that begs a question: how do bees understand the waggle dance in a dark hive?

To crack this quandary, studies began to focus on bees' ability to hear as well as see. It was discovered that they use something called a Johnston's organ to hear the sound of other bees' buzzing wings. Johnston's organs are small collections of sensory cells found in many insects' antennae. They vibrate tiny amounts when subjected to the movement of air particles, which, as with spiders' legs, means they tend to respond to near-field sounds. But once again, there is some evidence to suggest bees can hear further than just near-field. Some species of bees waggle their dance outside their hives, but for the ones that dance inside, darkness is only *one* of the difficulties in passing on information about one's next meal. The frequency produced by a honeybee's wings during flight is 250 Hz, which, believe it or not, is the note B (the first note of Chopin's beautiful étude *Tristesse*). But for the waggle dance, a bee's wing frequency is 256 Hz. Therein lies the next problem, in that this difference of 6 Hz is very small. It is what most trained Western musicians would perceive (in the midrange of musical notes) as around a quarter tone, a small differential of pitch around half the recognized minimum interval between two musical notes. So not only is the beehive dark, it's also full of noise from two competing and very similar sounding sources of buzz—flying bees at 250 Hz and dancing bees at 256 Hz. And just to make life even *more* difficult for those looking for the "buzzfeed" dancing signals, there are other bees fanning the hive to keep it at an optimum temperature.

To help bees tune in to that "buzzfeed" frequency, waggle dancers also use rhythmic bursts and intervals of silence to communicate more detailed directions—a kind of buzzing Morse code. Apiologists have found that honeybees are a little slow on the uptake when it comes to differentiating between flight and waggle dance, with up to a ten-second delay before the penny (or should that be beat?) drops. Until some form of pattern emerges, it can be initially difficult to tell in the dark whether a bee is flying in or out, fanning the hive, or performing a waggle dance. However, once the bees have realized that what they're listening to *is* a waggle dance, they cleverly move into an optimum position close to the dancer where the near-field air particle movement is most intense. Their Johnston's organs are sensitive to frequencies between approximately 200 and 500 Hz, which means that the waggle dance tone just above the note B is perfectly within their "hearing" range.

But just like any sophisticated form of communication, a variety of methods of messaging is key to delivering complex messages—wing frequencies and rhythmic bursts are not enough. Honeybees' legs, just like spiders' legs, are highly tuned to substrate-borne vibration. In the waggle phase, bees move their bodies in 15 Hz waggling motions, and it has been noted that more waggling occurs on open cells in the honeycomb rather than closed or capped cells. This suggests that bees can sense the resonant properties of the substrate they are dancing on. Along with the toots and quacks of queen bees and the stop signals of worker bees (all of which are also within the 200–500 Hz frequency range), that idyllic summer scene of a gently buzzing hive in a field of flowers could not be further from the truth. To a bee, it's a gigantic din, all within the confusingly narrow frequency range of the opening vocal lines of the Beach Boys' "Wild Honey." As it happens, Beach Boy Carl Wilson's first solo "Mama!" is a B, just around the upper limit of honeybees' hearing at 493 Hz. The lowest note of Wilson's phrase hits a G around 196 Hz, at the lower end of a bee's range (it's as if they planned it). However, researchers have yet to find video footage of any of the Beach Boys dancing in a figure eight, followed by the whole group running off to the nearest flower bed.

The Rat's Harp

At this point, you might be feeling more than a little depressed by the impact anthropogenic noise is having on those creatures with whom we share this planet. But not all animals are suffering. The following good news story is about a species that is thriving; indeed, it revels in our human world. This creature possesses a most ingenious frequency response system. To cheer you up, may I introduce...the rat. Perhaps one of the most commonplace vibro-sensory systems we would recognize is a whiskered nose on one of our pets. Most mammals, among them dogs, cats, gerbils, and rats, have whiskers. Vibrissae—which comes from the Latin *vibrare*, meaning "to vibrate"—is the official name for whiskers (and, in the medical world, for human nostril hair). They are thicker, longer, and more deeply embedded than normal hairs, specialized for tactile sensing and as acutely responsive to their owners as fingertips are to us. These whiskers are directly linked to the brain, to be precise, to the somatosensory cortex, an area dedicated to receiving and processing sensory information. They grow on various parts of the body, and beyond the usual snout area, they can appear in some rather unusual places—for example, on cats' front ankles (which are actually wrists, to be anatomically correct). The most studied of all whiskered animals, though, is the rat.

For many, the sight of a rat's beady black eyes, buck teeth, long twitching nose, and thick whiskers is enough to make their blood run cold. Its features conjure up filth, dirt, vermin-infested sewers, and the Black Death. It might help those suffering from musophobia (that's the technical term for a fear of rats, and not, as you might think, of musicians) to visualize their whiskers as a sort of beautiful musical harp, because in essence, that's what they are. Auditory neuroscientist Maria Geffen's first research project was to explore the resonant frequencies of rats' individual whiskers, in the unforgettably titled paper "Vibrissa Resonance as a Transduction Mechanism for Tactile Encoding." Geffen's team had two basic methods for testing the resonant frequency of rats' whiskers—in vivo (literally meaning "within the living") and ex vivo, where whisker plucking was required. What they discovered was rather startling and musically most intriguing. The general rule that longer musical strings produce lower notes (and shorter strings produce higher notes) also applies to rats' whiskers. As long ago as 1913, Northwestern University Medical School's S. B. Vincent published a paper in which he illustrated a grid of rats' whiskers, between five and nine columns labeled with numbers, and five rows (called arcs, because they curve) labeled with letters. Whisker A1 is at the top nearest a rat's eye, and E6 is at the bottom nearest its mouth.

Almost a century later, Maria Geffen used this grid to make a definitive acoustic map of rats' vibrissae. Geffen told me that she considers their whiskers to be like a "reverse piano," working in "frequency decomposition." If a rat wishes to explore something, it will "whisk" its whiskers back and forth at a rate of 10 Hz. The information that each individual whisker conveys to the brain means the rat can recognize the constituent parts of a set of complex frequencies. This is a superpower all musicians would love to possess: a talent to hear a complex chord and immediately recognize all its individual notes. Geffen said, "The reason why that's really cool is that it's an efficient code. It allows you to encode a complex signal with very little bits of information," which is naturally of great benefit to us humans in engineering terms. A musical example of this is our ability to compress data from studio recordings. Our mobile devices and cloud storage would be packed with large files, were we not able to compress a three-minute song from its original 3,500 MB down to its smaller 3.6 MB mp3 version. But of even greater musical value is that Geffen and her colleagues have led me to invent a new natural instrument, which I've named *vibrissae rattus*, or the rat's harp. As convincing as the Latin name might sound, though, it could yet prove difficult to persuade the angels at the pearly gates to toss their traditional instruments, leaving them to gently pluck the whiskers of rats nestled in their laps.

To find the resonant frequency of a whisker in vivo, Geffen's rats were first anesthetized. The unconscious subjects were laid on a floating table that minimized external vibrations, and each whisker was individually tested by sounding a large range of frequencies at short bursts of 500 milliseconds each. Infrared light was used to capture each whisker's vibrations, and the resonant peak frequencies were noted. To cross-check their findings, Geffen then removed the whiskers from the rodents and performed further experiments ex vivo. Rats' whiskers grow back relatively quickly, and research shows that their brains adapt to the loss of whiskers in the meantime, so it seems no harm was done.

But what of the *sound* of a rat's harp? Using the data from both experiments, the team produced a range of whisker resonant frequencies, from the lowest resonating at an 80 Hz E_2 (the opening bass note in the White Stripes' "Seven Nation Army") to whisker B_4 at 750 Hz (which is a smidge flatter than the pitch G_5, the first violin note at the start of Mozart's *Eine kleine Nachtmusik*). If one plucked both these vibrissae simultaneously, it could be claimed that rats are pitched in the key of E minor. The other whiskers cover a wide spectrum of pitches within this range, but none hit the precise frequencies we associate with our Western scale. A gentle brush of one's fingers across a rat's harp would be spookily ethereal, out of tune yet strangely compelling—a heady mix of serenity and sewer.

But what is the point of rats' individual whiskers having their own resonant frequencies? Could the answer lie partly in their environment: low light and the dark? As rats can be creatures of shadowy spaces, a supersensory tactile ability is essential. Rats cannot hear beneath 500 Hz, so their whiskers provide them with a vital extension of the frequency range they can access. In another study by neuroscientist Christopher Moore at MIT, the neurons in rats' brains fired with particular gusto when the whiskers vibrated around their resonant frequencies. The findings suggest that rats are aware of each of their individual whiskers, giving them a highly specialized neurological picture of their near-field surroundings.

But that's not the last of a rat's talents. In yet more studies on rats' whiskers (what is it about rats that gets scientists going?) at the McCormick School of Engineering, Professor Mitra Hartmann and her team discovered another important function that rats' vibrissae serve. They use their vibrating whiskers to calculate the direction of the wind, thereby allowing them to follow scents for food. This ability to help locate airflow sources—called anemotaxis—is a noncontact use of vibrissae that scientists had not really considered until recently. In experiments, the rodents were placed in a holding box before a small semicircular enclosed table; five equidistant fans blew air into their faces. In front of each fan was a rat-sized hole at the bottom of which was a liquid reward. Assuming whiskers play no defining role in the detection of airflow, the rats would mathematically have a 20 percent chance of picking the right hole. Over ten days, the rats consistently chose the correct fan and hole 55 percent of the time. Then, their whiskers were removed. Without the aid of vibrissae, there was a 20 percent drop in the rats' accuracy in identifying the direction of the wind source. Even when one considers that other sensory factors such as fur movement, skin, eyesight, smell, etc. might have some part to play in anemotaxis, vibration of whiskers through airflow seems to aid this sense, giving the rats an extrasensory string (or whisker) to their bows.

Not Only Corn Has Ears

Nature is full of vibration, and this chapter has focused on how vital frequency can be to the animal kingdom. But scientists are also recognizing that plants get hertz. Back in 1954, Indian scientists played violin renditions of Carnatic morning ragas to *Mimosa pudica*, a creeper of the pea/legume family. They claimed there was increased growth in spines, branches, and leaves of the plants. In 1963, unpublished data from a dubious-sounding Mr. G. E. Smith of Normal, Illinois (not to be confused with G. E. Smith, the guitarist who played with Bob Dylan and led the *Saturday Night Live* band), suggested that his corn and soybean seedlings liked George Gershwin's *Rhapsody in Blue*—but who doesn't? However, in the 1965 February issue of *BioScience*, Richard Klein and Pamela Edsall attempted to pour cold water on plants' aural and musical abilities. Under controlled conditions, they played an eclectic mix of music from five gramophone recordings: the Gregorian chant "Hodie Christus natus est"; the third movement of Mozart's Symphony no. 41; the Dave Brubeck Quartet's "Three to Get Ready"; "I Want to Hold Your Hand" by the Beatles; and, rather bizarrely, a rendition of "The Stripper." Klein and Edsall reported that "there was no significant effect of any of the musical selections on the growth or reproductive structures of the test plants." But they were all barking up the wrong plant, as it were.

A fixation on *music* as opposed to sound was at the heart of the problem. As we discovered in chapter D♭, all music is a mind-bogglingly vast, complex mix of frequencies, harmonics, and overtones, and the possibility that plants would react to such a frequency soup was always going to be very slim. In 1979, researchers focused their plant experiments not on music but on specific frequencies. Indeed, their chosen frequencies were in part dictated by the plants themselves. Using dwarf bush stringless green pod beans and impatiens as their subjects, they played them three randomly chosen pure tones: 500 Hz (around the note the trumpets play at the start of the famous bit of Gioachino Rossini's William Tell Overture); 5,000 Hz (one octave up into the Infinite Piano range); and 12,000 Hz (above the hearing range of many middle-aged people). The scientists wanted to explore whether a connection between leaf size and the wavelength of a frequency could produce an increase in growth rate. As mentioned in the black hole section of chapter A, wavelength is most easily calculated as the distance between two peaks of pressure. In air, each frequency has a specific wavelength; for example, the frequency of C_4 (middle C) has a wavelength of 1.301 meters.

The researchers found that there was an intriguing improvement in growth rate at 5,000 Hz and refined this frequency so that its wavelength equaled the bean leaf's six centimeters—around 5,600 Hz. They realized that these plants had a marked increase in growth rate when they were subjected to frequencies whose wavelengths were close to their leaves' dimensions. They suggested that such a wavelength "scrubbed" the leaf surface of any stagnant film of moisture, thereby helping the plant breathe more easily. On the face of it, this research should have exonerated Prince Charles, the Prince of Wales, who in 1986 was pilloried for admitting that he talked to his plants—a pastime enjoyed by innumerable horticulturalists over the centuries. However, it's not quite that simple. I've roughly calculated that the dominant frequency of the Prince of Wales's voice is approximately 165 Hz (which has a wavelength of 2.065 meters), meaning only a plant with leaves of gargantuan size—such as a banana plant—would be likely to respond to his regal words of encouragement.

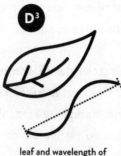

leaf and wavelength of
6 cm = 5,600 Hz

**improved growth rate
when wave and leaf
length match**

Many consider acoustic sensitivity in plants as just plain bonkers. But the potential is far from ridiculous. Humans have tympanic hearing (eardrums), but as this chapter has illustrated, this is by no means the only way of sensing frequency, vibration, and sound. The eminent bioacoustician Monica Gagliano argues that "birds and frogs have no outer ears, but their hearing can be more acute than ours." As substrate-borne vibrations are always present and pretty much everywhere, and hearing function can take a myriad of different forms, it is not too much of an imaginative leap to suggest that plants might exploit this rich resource of information. Gagliano, among others, argues that the time is ripe (excuse the pun) for more research focusing on plants' abilities to gather environmental information via frequencies that travel through ground or water. As both mediums are denser than air, vibrations pass through them much more efficiently. While most research has focused on the ways in which animals perceive sound and vibration, similar studies of plants have not been as advanced or recognized. A shrub just isn't as sexy as a snake, at least to the scientific community...and other snakes, of course. But the small amount of evidence that plants "hear" is both enticing and potentially far more revolutionary than animal-based research.

As this chapter closes, you might be wondering what became of our 52-hertz whale? Is it still out there desolately singing? Its last known calls were in 2016, though by then, 52's voice had dropped to around 46 Hz. I quizzed Joshua Zeman, the director of *The Loneliest Whale: The Search for 52*, a film of the expedition to seek out this oceanographic enigma. With a wide grin on his face, one of his opening lines in our conversation was, "I would say that finding one whale in the entire ocean is extremely difficult." He summed up the journey of the 52-hertz whale, saying, "It's a story that, in its most basic sense, creates the conundrum of the human condition—one whale calling out in the vastness of the sea." *The Loneliest Whale* has just been released at the time of writing, but did Zeman and his crew find the elusive 52? No spoiler from me, I'm afraid. His emotional journey to find the whale has led Zeman to understand the sentiment of those who wish 52's story to remain unsolved. "In a lot of ways, you don't want to demystify the magic that is nature," he acknowledged. But Zeman also noted, "The moment we started to listen, we started to hear things; we started to discover things we never knew existed." Doctor Dolittle was mistaken—we don't need to talk to the animals. We need to shut up and listen to them.

OF SOUND MIND

Our brains and frequency

One of my most memorable and rather nerve-racking gigs was a private recital to around twenty people, at which I premiered one of my own compositions. Almost half of the audience fell asleep. But far from a disaster, I was delighted, as the beginning of the gig had been dominated by a cacophony of crying, screaming, and wailing. The piece I had been commissioned to write was called "Sleep My Baby, Oh," and yes, half of the audience was under one year old. It was an experiment for the BBC, part of a film exploring the effectiveness of the lullaby. Although such babyish ditties are often described as "humble," not only can their impact be life-changing (to both baby and parent), but the science behind them is certainly more than child's play.

Following advice from an early years music expert, I set about composing four lines of melody that contained repetition, a strong falling phrase, and a calming resolution at the end—a "perfect cadence," in musical parlance. The beginning of the filming session was pandemonium: ten crying or agitated babies, ten undoubtedly sleep-deprived mothers (I'm not sure why there weren't any fathers present), and me, feeling a little foolish in front of them all, equipped with a laptop and piano keyboard balanced precariously on my knees. I was about to teach my new masterpiece to the mothers over the din of sobs and blubbering.

The first few minutes were truly awful. I appreciated how difficult it must have been for the mothers to learn a new song on camera while their babies bawled in their ears. I shared their pain, as I had a one-year-old at the time—we were all utterly exhausted. The only mild success was that the moms were gaining confidence learning the song, and I detected a small increase in their volume as the repetitions of the four lines quickly added up. However, the babies continued crying. But rather miraculously, after just five minutes, three of the babies were asleep, and by the end of the session around fifteen minutes later, all the babies were quiet, and seven out of the ten were snoozing contentedly. They might have fallen asleep anyway, perhaps from the sheer boredom of having a Mainwaring composition sung to them ad nauseam. Nevertheless, my lullaby seemed to have worked.

Often in filmmaking, one starts the cameras rolling with a clear view of the outcome, but I was genuinely surprised by the success of my tune (as were the mothers who were interviewed after the experiment). Though our methods were most unscientific and the whole project was hammed up for early evening light entertainment TV, it was undoubted the lullabies continue to work, just as they have for millennia. And there is growing neurological brain research to back up our intuitive faith in the power of music to affect even the youngest as well as strengthen the invisible bonds between the brains of mothers and their babies. Some of the most startling research in this area suggests that the brain waves of mothers and babies synchronize during the singing of lullabies, particularly when they gaze at each other. This sounds incredible, almost telepathic.

But what are brain waves? We all speak of having them during "Eureka!" moments, but in fact, we have brain waves all the time, even when we're asleep. It is estimated that our brains and nervous systems have around one hundred billion neurons, the brain cells responsible for sending and receiving tiny electrical signals that get us to do anything and everything. Neurons are a bit like trees, though considerably smaller. They have branches called dendrites, which receive messages from other cells, and roots called axons, which send information. And the trunk in the middle, the core of each neuron where the nucleus lies, is called the soma. Electrical messages—action potentials—are received by the dendrites and sent by the axons. For us to function in even the most basic of ways, our brain cells are in constant communication with one another, sending countless tiny electrical impulses throughout vast networks of neurons that synchronize in order for us to breathe, make a cup of tea, or perform brain surgery. These millions of synchronized messages are brain waves, and the rhythms of these electrical pulsations have been divided into five measurable categories, dependent on our level of brain activity and conscious state.

Using an electroencephalogram (EEG) test—which involves placing small electrodes all over a subject's head—clinical neurophysiologists can measure dominant brain wave states. If you were so comfortable during such a test that you fell asleep, your brain's electrical impulses would slow as the conscious world receded. The EEG machine would start to measure delta waves, electrical impulses that synchronize across your neurons between 0.5 and 3 Hz; these are the strongest waves in terms of amplitude or signal strength. As we begin to wake, our brain wave frequencies

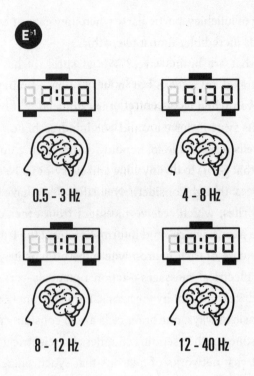

increase to between 4 and 8 Hz; we are still very relaxed and not particularly aware of the world around us. Impulses at these frequencies are called theta waves. At breakfast, during one of those early morning vacant stares out the window, our brain waves are mainly alpha types (now around 8–12 Hz), allowing us to relax with a passive attention to the world around us. By the time we might leave the house to go to work or to the shops, we are completely active and fully aware of the external world, our brains producing between 12 and 40 pulses per second, beta waves. And then, at those moments of the day when we need to fully concentrate, our brain waves can be firing at around 40 Hz or more—it's a wonder we can't hear them humming, as they are at the same frequency as the opening bass note of Cypress

Hill's "Insane in the Brain." Though these are the smallest waves in terms of amplitude, it's worth remembering that all our brain waves only produce electricity measuring just a few millionths of a volt. If shifted up eight octaves, our brain waves—from sleep to the moment of pure concentration in the office—would sound like a very extended version of the clarinet slide at the start of Gershwin's *Rhapsody in Blue*.

Let's get back to that strange psychic bond between mothers and babies. Recent international research suggests that they "show significant mutual neural coupling during social interactions"; in other words, their brain waves become synchronized. This 2017 research looked specifically at the theta and alpha bands, the waves produced when our brains are particularly relaxed. In a series of experiments, infants viewed their mothers singing nursery rhymes, in both live settings and via video, with a series of differing gaze positions: direct gaze, indirect gaze (head and eyes slightly averted), and direct oblique gaze (where the head is averted but the eyes look forward—if I was a baby, I think this would freak me out a little). Of all the tests, the researchers noted a significant influence on both mothers' and babies' brain waves when gazes were direct.

This backs up other evidence of synchronized brain wave activity among school children in class. Another 2017 report showed that shared attention in a classroom is the likely source of brain-to-brain synchrony, "'tuning' neural oscillations to the temporal structure of our surroundings." Likewise, studies of the synchrony of brain waves at live concerts confirm the strong links between brain waves, social interaction, and music. And as we pay attention in class or listen closely to a live band, we produce lots of beta waves around the same pitch as the bass Eb at the start of Stevie Wonder's "Superstition."

Mozart Makes You Smarter

Though my teenage listening repertoire broadened to include the likes of Frankie Goes to Hollywood, most of my musical favorites came from the classical genre. The music of Wolfgang Amadeus Mozart has been a companion of mine since I can remember, from the very earliest of my beginnings as a budding violinist at the age of five. I was therefore delighted to discover that my intelligence may have been enhanced by the countless hours I spent listening to Mozart during my formative years. Or this was what the witty and engaging music critic, journalist, and author Alex Ross wryly suggested in a 1994 *New York Times* article. He claimed that Beethoven was no longer the world's greatest composer—Mozart had usurped the grumpy genius through a newly discovered side effect. Ross reported that "researchers...have determined that listening to Mozart actually makes you smarter."

This throwaway comment created one of the greatest musical myths of the late twentieth century. Within no time at all, the "Mozart effect" was everywhere. Preschool kids were made to listen to the rondo from *Eine kleine Nachtmusik* while being force-fed nutrient-rich superfoods, and newborn babies were subjected to hour upon hour of Mozart's twelve variations based on "Twinkle, Twinkle, Little Star." Ironically, many readers of Ross's article stopped paying attention after "makes you smarter," missing the killer line at the end of the next paragraph, a sneering indictment of the Mozart-as-aural-wallpaper brigade: "If Mozart makes you smarter, why do Mostly Mozart audiences at the Lincoln Center break into applause in the middle of pieces, after slow movements as well as fast?" (Clapping in between the movements of a concerto or symphony is deemed to be a classic faux pas of the culturally illiterate.) But the Amadeus genie was out of the bottle, and the world went wild for the intelligence formula: developing brain + listening to Mozart = supersmart kids.

There was another fact missed by ambitious parents who were desperate not to raise stupid children. Ross's reference to a study by researchers from the Center for the Neurobiology of Learning and Memory never mentioned *all* of Mozart's music, or IQ, or intelligence, or even kids. The study was focused on a short burst of ten minutes of one single piece of Mozart, the Sonata for Two Pianos in D Major, and its impact on short-term spatial reasoning. In essence, if you listened to just a small section of this specific sonata (though which movement and which section is unclear), you could potentially picture more clearly what your sofa, chair, and TV would look like *before* you moved them to different parts of your sitting room...but you'd better do it quickly, as this improvement in spatial reasoning only lasted for around fifteen minutes, a sort of temporary boost of Mozart-induced feng shui.

But why the music of Mozart anyway? Throughout the twentieth century, Mozart's reputation as a wunderkind who played hard, died young, and had genius musical superpowers placed him in the "one of the best godlike great composers" category. Perhaps due to the ubiquitous nature and accessibility of Mozart, many therapists and scientists have used his work in their research programs, from sound therapist Dr. Alfred Tomatis to the 1993 research of Frances Rauscher and colleagues, leading to Don Campbell's 1997 book *The Mozart Effect: Tapping the Power of Music to Heal the Body, Strengthen the Mind, and Unlock the Creative Spirit*. But why not the music of Antonio Salieri? Even now, after years of experience, I would be hard pushed to differentiate between much of the two composers' music. And what makes the music of that specific era more stimulating to the mind than that which precedes or follows it? Mozart's spatial reasoning classic, the Sonata for Two Pianos, was written in 1781 in a style that was highly fashionable in its day. The trend for classical music (that written roughly between 1750 and 1825) was most definitely a reaction against the excesses of the preceding baroque style. The complex multilayered melodies, wall-to-wall florid decoration, and elaborate textures of the likes of J. S. Bach became—like all styles and trends—very passé, and Mozart, Salieri, and even J. S. Bach's own sons (such as C. P. E. Bach) wanted to keep up with the latest craze of catchy, simple, single-line melodies with predictably symmetrical phrase lengths and unfussy harmonies.

Could it be the lack of clutter in Mozart's music that gives us the ability to think more spatially? A paper published in the *Journal of the Royal Society of Medicine* by J. S. Jenkins in 2001 posed the question, "So, does the Mozart effect exist?" Jenkins noted that it couldn't be people's enjoyment of Mozart that added to their increased neurological performance, as rats completed maze tests quicker when listening to Mozart (though perhaps the rats enjoyed it too). Another piece of research analyzed a variety of music from the likes of Mozart, J. C. Bach, J. S. Bach, and Chopin as well as fifty-five other composers. In addition to spotting a regularity of phrase repetition patterns that was shared by Mozart and the Bachs, they found that there was "the emphasis on the average power of particular notes, notably G_3 (196 Hz), C_5 (523 Hz) and B_5 (987 Hz)." This is equivalent to saying that many of the paintings of Titian, Constable, and Picasso all emphasize certain shades of red, blue, and yellow. So what?

Perhaps the most undeniable evidence produced from the Mozart effect's heyday in the late 1990s and early 2000s was the high level of ignorance most people (especially journalists) revealed when discussing the basic inner workings of classical music. A report by ABC News in 2009 said that "according to researchers, the placid harmonies, sharps and flats, legatos and allegros of Mozart's music stimulate the brain, but relax the muscles." It is hard to overstate how preposterous the idea that Mozart's "sharps and flats" stimulate the brain, just as it's ridiculous to say that the commas and periods in Shakespeare's plays provoke our emotions. And what of the staccatos or andantes in Mozart's music; are they not as stimulating as the "legatos and allegros"? It was also claimed that monks from Brittany discovered that cows produced better milk yields when exposed to Mozart (or "Moozart," as the press inevitably dubbed him). In another report, microbes were apparently found to chomp through sewage more rapidly when Mozart was played through a sewage plant's PA system. (As a notorious scatologist, Wolfgang would surely have loved this.)

Though it's unlikely that French monks have made a significant impact on average milk yields, an Italian monk single-handedly changed the course of music history. Born around 991, Guido d'Arezzo was a music theorist and celebrated teacher who invented a rather clever music notation system. When Maria sings "Do, Re, Mi" in *The Sound of Music*, and François Truffaut encourages a keyboard player to perform "re, mi, do, do, so" to the aliens in *Close Encounters of the Third Kind*, they are both using a scale system that can be traced directly back to Guido. He claimed that his system enabled ecclesiastical singers to achieve, in one year, what would have previously taken them ten, and this seems eminently plausible. Once singers had learned his notation system, they could sight-read melodies instead of relying on rote learning and memory. His names for the medieval scale of six notes were as follows: *ut, re, mi, fa, sol*, and *la*. Each is the first syllable of the opening half-lines from the Latin hymn honoring John the Baptist, "Ut queant laxis":

> *Ut queant laxis,*
> *Resonáre fibris,*
> *Mira gestórum,*
> *Fámuli tuórum,*
> *Solve pollúti*
> *Lábii reátum,*
> *Sancte Joánnes.*

It is extraordinary that we have continued to notate music in this way for over a thousand years. We now know *ut* as *do*—a deer, a female deer—and *ti* was soon added, though not with jam and bread. As well as becoming the mainstay of musical notation for the whole of the Western world for over a millennium, Guido's solfège system has been making waves in more mystical frequency-based theories of late.

Defying Gravity

Dr. Joseph Puleo, one of America's leading herbalists, started exploring Guido's scale back in the 1970s. Though it was nigh on impossible to find any ancient references to specific frequencies for each of the notes of this scale (there was no way of precisely measuring a frequency before Savart's wheel of 1834), Puleo set about trying to "rediscover" the solfeggio frequencies, as they are called. He claimed to decipher a pattern of numerical codes in the Bible, specifically the book of Numbers, chapter 7, verses 12–83. Apparently through the use of the mathematical

Pythagorean Method, he concluded that the first frequency—
that of *ut* or *do*—is 396 Hz. This frequency is a fraction above
G_4, the high note Gnarls Barkley sings at the end of his first
phrase in the song "Crazy." All over the internet, it is claimed
that listening to this G (396 Hz) liberates guilt and fear. The
next note in the solfeggio frequency series is *re*, which resonates, according to Puleo, at 417 Hz, about a semitone above *do*.
This note, $G\sharp_4$ or $A\flat_4$, is the vocal first note (an octave lower)
in "My Girl" by Madness. Listening to *re* at 417 Hz is good for
"undoing situations and facilitating change." But *mi*, the third
solfeggio frequency, is undoubtedly the most important of
Puleo's discoveries. He claims it resonates at 528 Hz, a slightly
sharp C_5 (coincidentally, Crazy Frog sings a slightly sharp C at
the opening of his cover of "Axel F"). Many say that 528 Hz can
repair DNA and is also more generally responsible for transformation and miracles.

Leaving aside the claim that a single frequency can repair
broken DNA, undo situations and/or liberate guilt, the assembled scale of the solfeggio frequencies bears no relationship to
the one Guido was notating with his *ut, re, mi*. If Julie Andrews
had taught Puleo's solfeggio frequencies to the von Trapp children, they would have achieved more success marketing themselves as a comedy act. Just as humans do not possess an innate
ability to measure their running steps precisely to 4 miles per
hour, they are not capable of singing exact frequencies to within
1 Hz. There is, of course, the added question of how ancient
peoples measured frequency to find these special notes. Perhaps
they sang a slightly sharp $F\sharp_5$ *sol* at 741 Hz, which "awakens intu-
ition"? This is the first note played by the violin in the opening movement of Arnold Schoenberg's superb *Pierrot lunaire*, a
melodrama that explores lunacy among other things.

Frequency is often to be found resonating through another ancient mystery, the pyramids of Egypt. There is a frequency-based theory relating to their construction that, at face value, does seem rather far-fetched. Many suggest that the ancient Egyptians possessed strange, almost supernatural powers that enabled them to move giant stones via sonic levitation. And such mystical powers are not exclusive to this part of the world. In his 1961 book *Lost Technology*, Swedish engineer Henry Kjellson retells a story of one Dr. Jarl who witnessed Tibetan monks levitating large boulders during the construction of a monastery, using singing, chants, six trumpets, and thirteen drums of various sizes as their sound source. As well as drawing detailed sketches of the event, Dr. Jarl apparently filmed the levitation, but sadly the English Scientific Society (of which I cannot find any trace) confiscated all the films.

Such stories have attracted academic interest. Bruce Drinkwater, professor of ultrasonics at the University of Bristol, has produced some interesting calculations. He deduces that the volume needed to lift a 2.5-ton stone (similar to one of the numerous blocks used at the Khufu pyramid in Giza) would have to be greater than the loudest ever sound on earth, the eruption of Krakatoa in 1883, which is estimated to have produced 172 dB from 160 kilometers away. Moving a Khufu stone would require a sound volume of 187 dB. (To put this in context, you need to know two facts: an increase in 10 dB is equivalent to a doubling of perceived volume, and a jet take-off is at an eardrum-rupturing 150 dB from 25 meters away—so you'd potentially need a sound almost sixteen times as loud as a nearby jet engine!) In a fascinating conversation, Drinkwater told me, "The whole power of a power station is having to be fed into some loudspeaker that we don't have, all to levitate one

stone. It's a lot of effort to go to, isn't it?" He then raised the issue
of focusing the sound on to a single stone. Drinkwater's calculations seem to suggest that the wacky concept of acoustic levitation will never get off the ground...but what does a professor of
ultrasonics know about levitation anyway? As it turns out, quite
a lot. Indeed, Professor Drinkwater has constructed his own levitation machine. Granted, he cannot yet levitate 2.5-ton rocks,
but he has managed to suspend polystyrene balls in the air using
car parking sensors.

During our chat, I must confess that I was not expecting
the words *levitation* and *car parking sensors* to appear in the same
sentence, or even the same conversation. In a previous project in
which he explored how a computer could physically interact with
a human (called "haptic touch"), Drinkwater discovered that
parking sensors are perfect ultrasonic emitters. They are "dirt
cheap" and consistent in their construction due to mass production, and they generate sound at a safe ultrasonic frequency of
40 kHz, well above our 20 kHz audible threshold—this explains
why dolphins avoid parking lots. Although inaudible to us at
this frequency, the sound of a busy parking lot transposed down
seven octaves would be a reverberant racket of E♭s (around 312
Hz, the second note Paul McCartney sings on the word *long* in
"The Long and Winding Road"). Drinkwater told me that "each
one individually doesn't output that much sound, but they're
pretty high intensity actually. I think you can get something like
130 dB, which is really pretty loud...but when you connect them
together in an array with lots of them and focus the sound, then
you can get up to the 140 to 150 dB we need to achieve levitation." He continued (rather modestly, I feel), "This is one of my
main contributions I've made to the field of levitation, working
out that there are these cheaply available parking sensors you

can put together in an array, and using that approach, you can make really cheap levitators."

When two identical ultrasonic 40 kHz sound waves are fired at each other—one from above and one from below—they interfere as they interact, causing a series of nodes and antinodes. These are specific points of high or low intensity where the combination of the two waves either amplifies or negates air pressure. Drinkwater's levitating machine has a bowlful of upward-facing ultrasonic loudspeakers (parking sensors) and a second identical bowl facing down, with the combined energy of the sound waves focused on the center. In the low-pressure points where the waves interact (the nodes), Drinkwater and his team placed little polystyrene balls, which become trapped, defying gravity through levitation. The applications for such technology could be considerable, from tissue engineering to 3D displays. The professor explained how one of his projects has been manipulating cells using "acoustic radiation forces," very high frequencies in the megahertz range. "We were trying to pattern muscle cells on lines... It's like sprinkling seeds in your garden. You sprinkle them randomly normally, but the ultrasound would force them into lines, and because they were seeded in lines, you could grow better muscle fibers." And taking the principle of the levitating polystyrene ball one step further, Drinkwater showed how one could move a ball around at such high speed as to fool our brains into believing there was a 3D image before our eyes, just as one can appear to write in the air with a firework sparkler. Bearing all this in mind, it does seem highly improbable that Tibetan monks could levitate a stone 250 meters in the air with a few trumpets and drums. On the other hand, who would have thought you could levitate anything with two bowls of car parking sensors? So perhaps repairing broken DNA with 528 Hz is not that far-fetched? I'm off to repeatedly play a slightly sharp opening melody note of Maurice Ravel's *Boléro*, then I'll check my DNA.

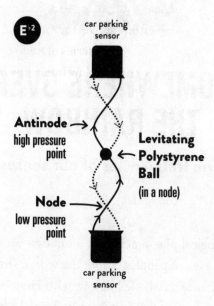

SOMEWHERE OVER THE RAINBOW

The frequencies of our senses

The meteorological phenomenon of a rainbow arcs through the legends and fables of peoples across the world, from the Maori rainbow god Uenuku (who fell in love with Hine-pukohu-rangi, the sky mist woman) to the Armenian myth of the belt of Tir, the god of knowledge. Real rainbows are as fantastical as the myths, just far more complicated. One begins with sunlight streaming from behind the observer, shining into rain or water droplets in front of them. As the light enters a circular droplet of water, it refracts (bends), splitting into its constituent colors. These colors are then reflected off the back of the droplet and bent again as they return out of the water, with a maximum intensity at an angle of between 40° and 42° to the original light beam. Due to the angle of the light and the position of the observer, the various layers of color are formed in the classic bow shape.

Up to now, we have predominantly explored *sound* waves. Such waves are mechanical: they need a medium through which to travel—air, water, etc. However, light is an *electromagnetic* wave. These waves do not need a medium to travel through, which is why light travels through the vacuum of space but sound does not. All matter has an electric charge—positive, negative, or neutral. Opposite charges attract, like charges repel each other, and these minuscule electrical shenanigans bring and hold together atoms. As they gain or lose charges through the transfer of electrons, an electric field is formed. When charged particles move, an electric current is generated, which simultaneously creates a magnetic field around it. In simple terms, these two fields can sustain each other, creating electromagnetic waves, or radiation as we might also know it.

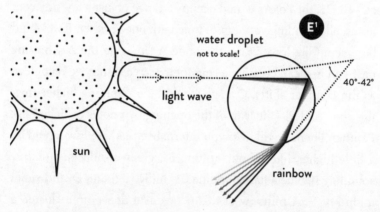

water droplet
not to scale!

E¹

light wave

40°-42°

sun

rainbow

Electromagnetic waves have amplitudes, wavelengths, and frequencies just like mechanical waves, but ears are useless for detecting them. We cannot hear electromagnetic waves; they do not change the pressure of air that vibrates our eardrums. However, we do possess an ability to see a very small part of the electromagnetic spectrum—visible light. If the light we can see was a set of mechanical wave frequencies, it would cover about one octave (twelve black and white notes) of the Infinite Piano, forty octaves or 6.7 meters above middle C. Beneath this octave lies infrared, microwaves, and radio waves. Above this octave, our Infinite Piano scale rises through the more harmful ultraviolet light at 10^{16} Hz (10,000,000,000,000,000 Hz), onward through octaves of X-rays, and up into gamma rays at 100,000,000,000,000,000,000,000 Hz and beyond (though it gets a bit more theoretical and fuzzy up there).

Transposing the frequencies of electromagnetic light waves down many octaves into the audible spectrum (and miraculously changing them into mechanical waves), a rainbow would surely sound ethereal and twinkly, right? Steel yourself to be disappointed. It depends on how many octaves we lower the waves' frequencies. If transposed down to our standard oboe tuning A at 440 Hz, the colors would occupy a range of just a few adjacent notes, all sounding very out of tune with one another. Red would be a semitone lower than the oboe A (imagine the A whizzing past you in a car, creating a semitone decrease that we recognize as the Doppler shift). Orange would be a semitone higher than the oboe A at B♭ (the leap of the opening two notes of "The Pink Panther Theme" will help you internalize this). Yellow would be a little higher, like an out-of-tune B♭; green would be a sharp-sounding B, blue a flat-sounding C♯, indigo a rough D, and violet a kind of E. A rainbow at this octave is a dense note cluster, a sound as musically sophisticated as a toddler's rainbow painting.

Fortunately for this chapter, one man from Edinburgh neatly tied together the areas of electromagnetism and color. His name was James Clerk Maxwell, and 1861 was a notable year for his work in both disciplines, as he presented a groundbreaking paper titled "On Physical Lines and Force" and also produced the first ever RGB (red, green, blue) color photograph. What is particularly profound about these two events is the impact both have had on our present-day lives. "On Physical Lines and Force" was the first time that the studies of electricity, magnetism, and light were truly married into one unified theory, which included four key equations that went on to underpin the work of such giants as Albert Einstein and still support much of today's information and communication technologies. On May 17, 1861, Maxwell presented, along with the esteemed early photographer and inventor Thomas Sutton, a lecture demonstration at the Royal Institution in London at which the first durable color photograph was taken. Maxwell developed the theory that full color photographs could be "assembled" using only red, green, and blue light.

The man responsible for much of the theory of color until then had been Sir Isaac Newton. In 1666, Newton began refracting light through glass prisms. By this stage in scientific understanding in the Western world, everyone was happily fixed on the idea of a five-colored rainbow—red, yellow, green, blue, and violet/purple-ish. By the time of his influential book *Opticks* of 1704, though, Newton had settled on *seven* colors of visible light—the red, orange, yellow, green, blue, indigo, and violet (ROYGBIV) we all know. This division of the spectrum into seven colors was in fact a rather arbitrary choice, based not on science but more on fad. For Newton, the problem with the five-colored rainbow was that it didn't neatly tie up with lots of other important universal patterns of seven, such as the days of the week, the planets of the solar system (as discovered at the time), and, most importantly, the musical notes of a scale. Newton wanted his rainbow to fit with these other patterns of seven. To dovetail with the Western musical scale, he took the accepted five colors of RYGBP (which isn't as memorable a mnemonic) and added orange and indigo, the latter being a very fashionable color in its time—indigo dye, imported from both India and the Americas, was one of the most expensive and exotic of commodities. Newton saw the musical scale as a spiral staircase, seven rising steps leading to the eighth exactly above the first, an octave higher. He imagined that the colors of the rainbow could follow a similar pattern—when one reached the top of the visible spectrum with violet, it would then transform itself back into red again.

Using this system, Newton should have started logically from the lowest frequency of the rainbow (confusingly the visual top!), which is red, represented by the musical note A. B would then follow with orange, C yellow, and so on. But on Newton's color circle illustration in *Opticks*, he labels red not with the musical note A but with a D. The reasoning behind this is due to the layout of a piano keyboard. It seems that Newton did not really regard his recently added colors of orange and indigo as worthy of full color status. The distance between D and E is a tone (it has a black note in between, D♯/E♭), whereas the distance between E and F is the smaller interval of a semitone (there is no note in between). Newton's red occupies the tone span between D and E, whereas the new shade on the block, orange, is only worthy of the semitone of E to F. Indigo fits perfectly into the *other* semitone gap, between B and C. The result of all these colorful musical capers is that Newton's annotated rainbow scale, starting on red and rising through the frequencies of the visible spectrum, is a type of scale neither major nor minor—it is a modal scale, specifically the Dorian mode. Though it might have been familiar to Newton through popular songs of the day, it was hardly used in the classical music of the eighteenth and nineteenth centuries. Returning to fashion in twentieth-century jazz and pop music, it is particularly striking in pieces such as Miles Davis's "So What," Dave Brubeck's "Take Five," and, rather appropriately, Pink Floyd's "Any Colour You Like" from *The Dark Side of the Moon*.

Building on Newton's ideas during the nineteenth century, Thomas Young—and later Hermann von Helmholtz—led science to the theory of trichromatic color vision. These two discovered that we perceive colors through three slightly different types of receptors in our retinas, called cone cells. These cones are "tuned" to receive different wavelengths of light, from the longer wavelengths of red, the middling wavelengths of green, to the shortest of the three, blue. Red has a wavelength around 700 nanometers (a nanometer [nm] is one billionth of a meter; a wonderful description of one nanometer is the distance an aircraft carrier would further sink into the water if a seagull landed on it), and its frequency is around 455,000,000,000,000 Hz (455 terahertz). Green's wavelength is about 540 nm and its frequency is 560 THz. And blue's wavelength is 460 nm, its frequency approximately 600 THz. These three colors' frequencies—the primary colors of light—transposed down forty octaves would sound like the opening melody of Booker T. and the MG's greatest hit, "Green Onions"—which should henceforth be renamed "Red, Green, Blue Onions."

The observant reader will have noticed the inverse relationship between wavelength and frequency—as one increases, the other decreases. Perhaps at this point, it is worth taking an illuminating digression into the relationship between wavelength, frequency, and speed, through the obvious medium of surfing.

Clad in a wetsuit, running down the beach to an accompaniment of "Good Vibrations," you squint toward the ocean, poised to catch your first electromagnetic light wave. Disappointment quickly ensues through the realization that electromagnetic waves travel at only one speed. You cannot catch a quick wave or a slow wave; they all travel at a constant velocity. The good news, though, is that light waves travel at 300,000,000 meters per second, so it's going to be one hell of a ride. The second thing you need to know is how often these waves will appear while you sit on your surfboard, waiting to catch a light wave. If the waves' frequency is quite low—in other words, you spend more time waiting than surfing—then the physical length between each wave will be large. However, if the waves are very frequent, the distance between the top of each wave will be very small—the wavelength will have reduced and the waves' frequency increased. But in the excitement, you must keep reminding yourself that no matter how many waves there are or whatever the distance between each peak, the surfboard will only travel at one speed, 300,000,000 meters per second—quite a ride!

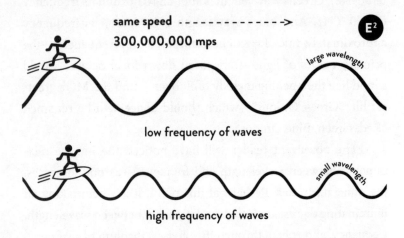

The Young-Helmholtz theory that our eyes perceive the full color spectrum through a variety of blends of only three primary colors led the Scottish mathematician and physicist James Clerk Maxwell to that practical demonstration at the Royal Institution in 1861. His photographic subject was a little tartan rosette, chosen for its broad range of colors and undoubtedly its Scottishness. Three identical images were taken through red, green, and blue filters. These images were then projected back through the same filters, superimposed on top of each other, and voilà...a color picture. Unfortunately, his photographic guru, Thomas Sutton, had not quite perfected the chemical sensitivity of the photographic plates, leaving them insensitive to red light and not particularly useful with green either. However, Maxwell and Sutton got lucky as the plates were, unbeknownst to them, sensitive to ultraviolet light, meaning the image was photographed via a little green, blue, and ultraviolet. Maxwell commented on the lack of success during the lecture, but nevertheless color photography as we know it was born.

The Taste of Vibration

As well as sight, other human senses are also affected by frequency and vibration. For those who like a bit of spice in their lives, hertz have a hand in cooking up the "buzz" of some of the world's tastiest meals. Famed for its distinctive and fiery cuisine, the Szechuan province in southwest China boasts a rather unique cooking ingredient, Szechuan pepper. It is not really a pepper that we would recognize, though, close to neither black nor chili peppers. It comes from the genus *Zanthoxylum* or prickly ash, and the variety of small trees from which the peppercorns are harvested is suited to the warm and damp subtropical climate of Szechuan. Many would argue that Szechuan pepper is the defining ingredient in the region's cuisine, its worldwide reach now firmly established through dishes such as kung pao chicken. The spice itself is not particularly hot, but when consumed or even merely brushed on the lips, it induces a tingling and numbing sensation.

In 2013, Nobuhiro Hagura (among others) investigated the reasons why Szechuan pepper induces tingling on the lips and tongue. Through four different experiments, Hagura and colleagues discovered that touch receptors were responsible for generating the tingle, and interestingly, the twenty-eight participants in the study led the researchers to the conclusion that "the frequency of the tingling sensations was strikingly consistent." Ground Szechuan pepper was applied to the subjects' lower lips, and once the tingling began, a tiny mechanical vibrator was placed on their right index fingers. Though a large range of random frequencies was generated through the finger vibrator, the subjects all agreed that a frequency around 50 Hz produced the most similar tingling sensation on their fingers to that created by the pepper on their lips. The experiment suggested that Szechuan pepper generates a sensation like placing one's mouth on a double bass's lowest string as it plays the opening note of Elgar's Introduction and Allegro for Strings, a G_1 at 49 Hz.

Our sense of touch and pressure is defined by a series of sensory fibers and four main types of mechanoreceptors, each specializing in different speeds of response and frequency. They have glamorous names such as Meissner's corpuscles, Ruffini endings, and Merkel's disks (though this sounds more like a radio show hosted by the former German chancellor). Meissner's corpuscles are situated close to the surface of our skin, have very small receptive fields, and work most efficiently in the 10–50 Hz frequency range (perfect for the low organ notes in César Franck's Choral no. 3). They are concentrated in areas where a sense of light touch is important, such as the fingers and, yes, the lips. However, sustained stimulation leads to Meissner's corpuscles losing their receptive abilities.

In a further experiment, the researchers used a psychological method called "adaptation," inducing a fatigue in the sensory fibers. They applied the 50 Hz vibrating machine to the subjects' lips over a minute or so and then applied the Szechuan pepper. What they noted was that "the perceived frequency of pepper-induced tingle was significantly reduced when preceded by an adapting mechanical vibration on the lips." In other words, the frequency of the pepper decreased after they had placed the vibrator on the subjects' lips, and the number of pulses induced by the pepper was reduced. Another interesting discovery that Hagura told me about was that the tingling sensation on the lips disappeared when pressure was placed on them—and it returned when the pressure was released. This possibly suggests that our tactile systems can inhibit and override different means of feeling. Instead of our systems receiving information simultaneously from a broad range of stimuli such as vibration, temperature, texture, etc., it seems we have the potential to be able to switch off individual fibers and receptors. Could this provide opportunities in the future to bypass damaged parts of our tactile system?

This research was widely reported throughout the media in 2013, and as is often the case, after initial mild curiosity, many asked why good money was being spent on such trivial nonsense— didn't scientists have better things to do than find the frequency of Szechuan pepper? Should all who eat such cuisine now carry a small vibrator with them to place on their lips just before their food is served, helping reduce the tingling sensation? What was missed by the popular media, though, were the gains made in our understanding of the human sensory system. First, the research report suggested that "decomposing complex somatosensations into component units of neuronal activity is a critical step towards understanding how the brain constructs sensory experiences." In simpler English, such work is vital if we are to understand how the brain helps us sense things. But of more precise value is that Hagura's work could have an impact on our treatment of more chronic forms of paresthesia. From time to time, we all suffer from a mild version of this—pins and needles. However, chronic paresthesia can be a much more debilitating condition, a symptom of nerve damage or an underlying neurological problem. The National Institute of Neurological Disorders and Stroke says that paresthesia "can be caused by disorders affecting the central nervous system, such as stroke…multiple sclerosis, transverse myelitis, and encephalitis." Research proving that Szechuan pepper makes your lips tingle at 50 Hz could ultimately be of great benefit to sufferers of chronic paresthesia.

Feeling the Vibes

Though there are four main types of mechanoreceptors under our skin, these are not the only ones. There are also less specialized unencapsulated free nerve endings, which detect pain, itches, etc. These are the worker bees of our somatosensory system: there are lots of them, they don't get a fancy name, and they do the slightly boring stuff. But there is another set of rather enigmatic, some would say glamorous and titillating, mechanoreceptors that can be found in an area many scientists have been too coy to take a peek at. One man who dived right in was German anatomist Wilhelm Krause, after whom is named the end bulb of Krause. If you think that sounds a bit suggestive, you'd be right. As well as the lips and tongue (and a bit randomly, the eye), Krause end bulbs are specialized receptors that can be found in both male and female genitalia. And the link between vibration and genitalia is as strong as the connection between trains and tunnels or corks and champagne. Krause end bulb receptors (also known as bulboid corpuscles or genital corpuscles) add a little je ne sais quoi to pleasurable touch.

Surprisingly, perhaps, male and especially female genitalia—as well as sexual function—are areas of science that have not been particularly well researched. We do know that the glans

penis and the glans clitoris are packed with all kinds of receptors, which are especially dense in the clitoris. Over the last few
decades, manufacturers have jumped on this corpuscle band-
wagon, and the internet is now awash with products with names
like Magic Wands, Rabbits, and Smile Makers. Those of a more
prudish nature will be mortified by the explosion in little vibrating toys, undoubtedly warning that the end of the civilized world
is at hand.

However, the penultimate sentence is not quite true.
According to a startling Cambridge University Press article by sex
historian Hallie Lieberman, "the vibrator was pervasive in con-
sumer advertising" during the first three decades of the twentieth
century. From posters on streetcars, through ads in the *New York
Times*, to displays in electrical shop windows, vibrators were tar-
geted at everyone. Yes, everyone: from grannies to infants, "self-
made" men to stay-at-home mothers. Vibrators were "health"
products, and their ads were sneaked under the censors' radars
as such. They could be used to massage one's neck, arms, legs, or
even be utilized in the home as a household appliance, a "labor-
saving" device. Manufacturers often did not specify what vibrators did that was particularly "labor-saving" (soothing a colicky
baby or stimulating hair growth in bald men, apparently), but
consumers well understood the message.

In an entertaining YouTube video posted by Callaly titled
"Unboxing 1930s Women's Products," the film's participants
explore an antique vibrator. Its pitch is about 196 Hz, spot on
a G. As everything in the world seems to accelerate, counterin-
tuitively, the frequency of vibrators might have *decreased* in the
succeeding years. Contemporary vibrators have a range between
around 40 and 150 Hz. On a bass guitar, this range encom-
passes the opening E_1 at the start of Frankie Goes to Hollywood's

"Relax" up to the high E♭$_3$ that opens Charles Wright's "Express Yourself"—musical proof, if any were needed, of the range of the most stimulating frequencies.

As stated earlier, receptors have a number of different functions—Meissner's corpuscles are fast-acting vibration sensors at frequencies between 10 Hz and 50 Hz, and Merkel's disks have a peak sensitivity up to 40Hz, detecting and, importantly, *remaining* sensitive to light pressure. But Pacinian corpuscles, reactive to even the slightest of vibrations (similar to the ones in Ning Nong the elephant's feet), appear to filter out all lower frequencies under 250–350 Hz. This seems to suggest that vibrators that function above 50 Hz are not particularly effective, yet a Turkish study in 2015 indicated quite the opposite. Thirty-two women were divided into five predetermined groups, depending on the stimulation frequency at which they could attain clitoral orgasms: 87 percent of the women said the vibrators made their orgasms easier; 85 percent said their sexual lifestyles were improved; and 79 percent said the bullet vibes improved their libido. Such studies might seem a little vague to scientists—which is to be expected—as orgasm is as much about brain wave frequency as it is about mechanoreceptor frequency. But research into this area also takes a more practical approach, especially regarding sexual health and disorders such as anorgasmia and erectile dysfunction.

The Fragrance of Frequency

An argument between opposing theories of another human sense has been creating quite a stink. Our sense of smell—olfaction, as it is known in the trade—is yet to be fully understood (not that we fully understand any of our senses). We know how smell gets from the receptors in our noses to our brains, via two olfactory bulbs. Indeed, the olfactory system is the only part of the brain that protrudes through our skulls. We know that smell, along with taste, is driven chemically and not through waves, the primary source for our senses of vision and hearing. But that's where the arguments start—some suggest that waves and vibrations *do* play a part in our ability to smell. The two opposing camps have starkly differing principles at the core of their smelly theories. The chemical camp states that odor molecules bind to receptors in our noses using the lock and key method, in which the specific shape and size of a given molecule fits into a specific receptor (we have around four hundred to five hundred in our noses), leaving our brains to interpret the receptors' signals as the smell of brewed coffee, freshly cut grass, or lemons. The vibration camp argues that as molecules vibrate, our receptors "listen" to these vibrations, which our brains translate into wet dog, burning plastic, or toasted bread. The arguments between the opposing theories have become quite public and rather heated. The big guns who have recently been slugging it out in academic journals have been headed up by Eric Block for the chemical team and Luca Turin for the vibration team. What has been fascinating to watch is how some of the published papers have not held back in their criticism of the opposition. It is easy to imagine Block et al. holding their noses while typing the title of their 2015 paper "Implausibility of the Vibrational Theory of Olfaction." During the same year, in the same journal, Leslie B. Vosshall weighed into the argument with "Laying a Controversial Smell Theory to Rest." Luca Turin's vibrational theory of olfaction really does seem to have gotten under the skin of many leading chemistry and biology experts.

Turin resurrected and developed this startling idea that our noses "listen" to vibrations, first postulated by Dr. Malcolm

Dyson in 1938. It is common scientific knowledge that all molecules vibrate (a molecule is a group of two or more atoms that have joined together). In simple terms, when atoms create a molecule, they are joined via bonds that we can imagine as pieces of elastic or springs. The atoms bounce around, vibrating at the ends of the bonds in rocking, wagging, or twisting motions, and the speed at which they vibrate has been measured at between 10^{13} to 10^{14} Hz (10,000,000,000,000 Hz to 100,000,000,000,000 Hz). Within this range, a molecule's frequency depends on its makeup and the strength of its bonds. Water is made up of a hydrogen atom, an oxygen atom, and another hydrogen atom—hence H_2O. In a water molecule, an oxygen atom has only six electrons in its outer shell and thus is two electrons short of eight, the stable configuration that it seeks, so it shares one from each of the hydrogen atoms (which are also one short of a full outer shell). This way, all are happy and the molecule is held together by these covalent bonds. As an example of a molecular resonant frequency, nitrous oxide (N_2O, laughing gas) is 5.63×10^{13} Hz, which means that the bonds that join the atoms vibrate at 56,300,000,000,000 Hz. This frequency, thirty-eight octaves lower (around 6.3 meters on the Infinite Piano), sounds a slightly flat A♭ at 205 Hz (Andy Williams's opening note of the song "Moon River").

Luca Turin cites five atoms as the basis for all smells we encounter: carbon, nitrogen, hydrogen, oxygen, and sulfur. Part of his argument is that, if the lock and key theory of olfaction is correct, two molecules that have the same shape—for example, cis-3-hexenol and cis-3-hexenethiol—should smell the same. But they do not. The former has an alcohol group on the end and smells of cut grass, whereas the latter has a thiol at its end (which does not change the shape) and smells of rotten eggs. Turin claims that if the lock and key method is solely responsible for our sense of smell, an olfactory receptor would not be able to identify the difference between the two molecules with the same shape, meaning vodka could potentially smell of rotten eggs. However, the molecular vibrations of *cis*-3-hexenol and *cis*-3-hexenethiol are different, suggesting that receptors sense the smell of cut grass or rotten eggs through the frequencies of the vibrating bonds in each molecule. In a recent BBC documentary on this subject, Professor Jim Al-Khalili used the case of almonds and cyanide as a convincing argument to back up the molecular vibration theory. They have different molecular shapes but share the same vibrational frequencies—and they have a similar smell. He suggested that this supports the theory of vibrational olfaction, which has recently been given the rather sexy term *quantum biology*.

This is all very theoretical, but it becomes interesting to the layperson (particularly the style-conscious one) when applied to one of the world's most lucrative and glamorous of smelly obsessions: perfume. Luca Turin straddles both the world of science and the art of perfumery, a feat rarely achieved. But there is also a third connection, that of music. The craft of a perfumer is a creative process dominated not only by oils, smells, and a blend of art and chemistry but also by time and duration, not unlike that of a composer. A perfume's smell is timed to "play" different notes to its audience, from the initial spray burst until the smell dies away or is washed off. Turin said, "Roughly speaking, every time you add one carbon to an odorant, you delay the evaporation by a factor of two." This allows the perfumer to time the introduction of different constituent parts of their creation. The basic smell structure of a perfume consists of three distinct "notes": a top note, which Turin said can typically last for about ten to fifteen minutes; a heart or middle note, which will then dominate for the next few hours; and a third note called a base note or "dry down." In a wonderfully fascinating conversation with me, he explained that the classic feminine fragrances of yesteryear had a top note that was less important, like an overture before the curtain is raised on the main show, leading the audience toward the high point an hour into the heart note and leaving a quiet and lingering dry down at the end. Turin has observed that more recently, to appeal to people's current buying habits, perfumes are front-loaded with a very powerful top note that often lasts for only a few minutes. As many people buy perfumes in a store, it makes commercial sense to give the purchaser an instant bang for their buck if you want to sell a fragrance.

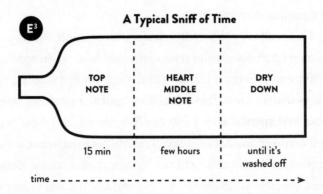

Of course, each of these "notes" can be a complex blend of molecules in themselves. What gives these molecules commonality is their volatility, the speed at which they evaporate. Turin explained that a typical top note might be lemon oil, which could contain three or four main aroma chemicals, each of which could be a ten-carbon alcohol producing at least twenty-four atoms, giving the note over sixty individual molecular vibrations. Pushing the top of a tester fragrance in a department store might be triggering our noses and brains to interpret a complex series of quantum frequency calculations. If Turin's theory is true, our olfactory receptors are bombarded with a dense chord, a musical feast of over sixty different waves, from which our brains not only exclaim, "Ooh, that's so lemony," but also subjectively decide how attractive the smell is. Even then, the sixty-note lemon chord has not yet finished stimulating our minds. The amygdala part of the brain is immediately set to work putting it into a personal context—the smell might remind us of another person, a specific time, place, event, color, or, in Turin's case, a musical interval.

In my conversation with Turin, he revealed another link between perfumery and music. Chemical compounds called esters are the result of the combination of an acid and an alcohol, and Turin states that the vibrational ratio between the two creates, in his mind, a perfect fifth musical interval (there is a strong 3:2 ratio at work here). This ester linkage is like a special chemical Lego piece to which one can attach any other piece on either side, which is why there are thousands of different esters in perfumery. But that special linking piece of synesthetic Lego "is a fruitiness I perceive as harmonically correct," stated Turin. "I always consider esters to be Mozartian." However, he added, "This is nothing more than idle fantasies of a perfume fanatic!" But the link between smells, our noses, and our brains is an incredibly complex one, a blended aroma with a top note of chemistry and quantum biology, a heart note of artistic perfumery, and a dry down of individual psychology.

Whole Body Hearing

Our own bodies are constantly bombarded with frequencies, the interpretation of which allows us to see, taste, possibly smell, and definitely hear. But to treat our sensory tools in isolation is simplistic, leading us to miss a much more holistic understanding of how we perceive the world around us. For example, we don't only taste with our mouths; we need our sense of smell (and some would claim our eyes as well) to create a true taste sensation. Similarly, with our hearing, we do not exclusively use our ears.

Dame Evelyn Glennie is the world's premier solo percussionist. For most, that would be achievement enough, but as Glennie was considered profoundly deaf by the age of twelve, this makes her achievement truly astounding and totally inspirational. For those with some form of hearing loss, Glennie has rewritten the rulebook on musical study. Indeed, she has inspired generations of musicians of all backgrounds, who might previously have been discouraged from finding ways to overcome both physiological challenges and societal obstacles. Glennie might not hear music as most people do, but she certainly feels it. We all feel music, but more often than not, we undervalue and even ignore the vital part of our auditory system that isn't our ears—namely, the rest of our bodies.

As a young music student, Glennie learned to focus on the vibrations and frequencies that resonate through our bodies whenever we listen to a live musical instrument. In conversation, I asked about her ability to tune timpani (kettle drums, which have specific pitches) without being able to hear them. A standard twenty-six-inch timpani has a range between A_2 and F_3 (110 Hz and 174.61 Hz), the large leap at the start of the melody of Scott Joplin's rag "The Entertainer." Glennie's first percussion teacher, Ron Forbes, led her to begin to feel the vibrations of a timpani by putting her hands on the wall of the music room. During her early secondary school career, Glennie discovered she could no longer hear the duration of such sounds. She could feel the impact of the drum being struck, but there was no tail to it. Forbes told her that as the drum vibrated, so did she. He enabled the young Glennie to begin focusing on where in her body she felt vibrations and, equally, to try to follow the journey of that sound as it decayed. Forbes then struck a second, different pitched timpani and asked her to detect in her body where she was feeling those vibrations. At this stage, Glennie could not necessarily pinpoint a location, but she could feel a difference.

Incrementally, Forbes lessened the frequency gap between the two pitches, allowing Glennie to be able to discern different notes. She says the higher the notes, the less she could feel them. As the "Ghost Notes" chapter reveals, our bodies' physiological makeup seems to sense more of the low end of the spectrum, including frequencies outside our own audible range. An ability to play with flawless technical skill and incredibly keen "hearing" are not, though, the only skills required by a leading musician. One of Glennie's attributes that is truly remarkable is her laser-like focus on the "journey" of each and every note and their consequent impact on her audience. Even though she has good reason to focus on herself, Glennie always puts the audience first. She is unceasing in her desire to interpret and deliver music for the benefit and joy of others, in order that they may "feel" the music as well as hear it.

But one area where Glennie is just like us mere mortal musicians is that she is suffering from presbycusis, the process whereby we lose the ability to hear higher frequencies with age. She said, "If I'm sitting at a drumkit, I find myself almost as though I'm twelve years old again, pounding the hi-hat." This frequency degeneration can be due to different factors, including age, exposure to noise, genetics, and medicine intake. Presbycusis seems to be an inevitable and natural process, but a notable study in 1962 found that the Mabaan tribe of the quiet Sudanese desert retained their hearing into old age. Though the study throws a little doubt on the inevitability of presbycusis, it is hard to determine whether it was the Mabaans' relatively peaceful environment or their genetic makeup that was the main factor for their extremely good auditory health. Glennie describes her experience of presbycusis: "Glockenspiels, cymbals, tubular bells, triangles I could hear very well, but because they were all mushed in together, I couldn't work out how to differentiate the feeling. That's taken a long time, so now I'm beginning to work that out. Just as I'm learning to feel it, the sounds are disappearing."

The main area of study of presbycusis is focused in and around the cochlea. This startling little snail-shaped miracle, measuring only around nine millimeters by five millimeters, is a phenomenal piece of acoustic engineering. But what does it do, and how does it enable us to hear?

On May 5, 1961, Alan B. Shepard blasted off in *Freedom 7*—the Mercury capsule atop a Redstone rocket—and thereby launched his astronaut career with the notable achievement of being the first American in space. However, his future prospects were soon in free fall as Shepard started suffering from Ménière's disease, an affliction of the cochlea whose symptoms include dizziness, tinnitus, and vomiting. Step forward one Dr. William F. House, who cured Shepard in 1968 with some rather experimental surgery on the astronaut's left inner ear. Up until this point, Shepard's career was dead in the water, but House's dicey surgical intervention allowed him to go on to command the Apollo 14 lunar mission in 1971. Arguably, though, this was not the biggest achievement of House's professional life. In 1961, exactly a hundred years after Dr. Prosper Ménière first posited an explanation for the debilitating bouts of vertigo and nausea—and coincidentally the same year Shepard blasted off from Cape Canaveral—Dr. William F. House fitted the world's first true cochlear implant.

In essence, the implant is a bypass road around the outer and middle ear. A small type of microphone attached around the ear translates sounds into electrical signals that are then fed through a tiny wire directly into the cochlea. At the other end, the signal bypass rejoins the normal auditory route from cochlea to brain. However, much controversy still surrounds the use of cochlear implants. They are not hearing aids and do not restore hearing. Cochlear implantees can gain some sense of sound and hearing but often need a sustained amount of ongoing therapy to train their brains to interpret the messages they are receiving from the cochlea. Much of the debate is around the erosion of Deaf culture and the devaluing and downgrading of sign language, particularly among young children whose parents have opted for the procedure. But why did House decide to rejoin the auditory path at the cochlea? Why did he not send the signals directly to the brain? The answer lies in the extraordinary work of this thirty-five-millimeter-long frequency miracle.

The Extraordinary Cochlea

The cochlea is part of the mystically named bony labyrinth. The whole structure controls our balance at one end and our hearing at the other. The cochlea part resembles a very small snail shell, coiling in on itself approximately two and a half times. Imagine that the spiral is one of those multistory parking structure entrances, ascending floor after floor in a seemingly never-ending right or left turn; sound waves are the vehicles entering our cochlear parking structure. Ingeniously, the road is spring-loaded, fitted with underfloor sensors that indicate when a car is on the ramp. Even more ingeniously, this spring-loaded road has a complete range of incremental weight-specific sensors—the start of the road is only sensitive to the weight of light bikes; by the middle, sensitivity is fine-tuned to heavier car weights; and by the end of the road, the sensors are filtering out all but the heavy trucks. The person in the control room of the parking structure knows how many vehicles are entering and what type of vehicles they are. This is exactly what the cochlea does, except that its sensitivity is triggered by frequency rather than tonnage.

Within this tiny thirty-five-millimeter spiral, hair cells are triggered by frequencies from 20,000 Hz at the opening all the way down to 20 Hz at the center. A frequency of 5,000 Hz (recognizable to many of us with tinnitus) will have a peak hair cell sensitivity around ten millimeters along the cochlea. Trumpeter Maynard Ferguson's final A_6 scream of a note on his recording of the theme from *Rocky*—around 1760 Hz—will fire up nerve impulses around seventeen millimeters along the spiral. The middle C (261.63 Hz) that starts the famous Andante of Haydn's "Surprise" Symphony will set hair cells a-tingling around twenty-seven millimeters; and Concorde's sonic boom, which might have thrown Whitetail the pigeon off course in chapter B, will stimulate nerve cells from around thirty millimeters to the end. The name *cochlea* comes from its shape, the ancient Greek word for "spiral" or "snail shell"... which is rather ironic, as snails can't hear.

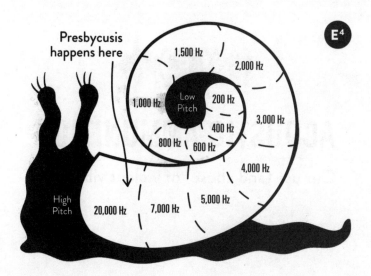

"Cochlea" the snail

Vibrations and frequency seem to be central to all our major human senses, from the 50 Hz taste of Szechuan pepper to the 455,000,000,000,000 Hz frequency we see as red light. However, the concept of five senses is a very outdated one. There are now competing seven, nine, and eleven sense models, with some even suggesting up to twenty senses being possible. Newton definitely hedged his bets correctly when he added orange and indigo to increase his rainbow to seven colors— it makes perfect sense to reflect the seven days of the week, the seven wonders of the world, seven musical letters, seven musical sharps and flats, etc. But nine, eleven, or even twenty colors of the rainbow? Please let it not be true that the lyrics of Andrew Lloyd Webber and Tim Rice's "Joseph's Coat" are revealed to be more scientifically accurate than the workings of Sir Isaac Newton.

ACOUSTIC AMMUNITION

Our use (and abuse) of violent vibrations

One of the most bizarre tales in horse racing history occurred during the 5:30 at Royal Ascot on June 16, 1988. The King George V Handicap, a flat horse race run over 1 mile, 3 furlongs, and 211 yards (2,406 meters), offered a first prize of over £11,000. Eighteen jockeys and their horses were eager for the start, which was almost nine minutes after the advertised start time. The favorite was Ile de Chypre, a three-year-old British thoroughbred whose career was in the ascendancy—his odds were 4:1. His trainer, Guy Harwood, was delighted to see Ile de Chypre leading from the front, and with experienced jockey Greville Starkey in the saddle, the race looked all but won. Still comfortably ahead in the final furlong, Ile de Chypre suddenly swerved violently to the left, throwing Starkey to the ground. Second-placed Tony Culhane and his horse Thethingaboutitis couldn't believe their luck and romped home to victory. Ile de Chypre trotted riderless over the line while Starkey dusted himself off, 140 meters farther back.

This, though, wasn't the bizarre part of the incident. Horses often get spooked, or jinked, as it is known in the world of horse racing. Many years later, legendary jockey Willie Carson, who had also ridden Ile de Chypre, remembered him as "a very tricky horse" who could be prone to this type of temperamental reaction. Indeed, the incident was so un-newsworthy that it passed with but the merest of comments in racing circles. In November 1989, though, over a year after the event, Ile de Chypre's jink was thrust into the headlines after it was used as evidence in a trial at Southwark Crown Court in London.

Car dealer James Laming, along with three others, was charged with a variety of drug-related offenses. Banknotes with traces of cocaine were found in his car, and links with Peruvian drug gangs were soon being investigated. Laming's defense included the startling revelation that he had "nobbled" Ile de Chypre with an ultrasonic stun gun of his own invention. Carrying any gun into a horse race is generally frowned on, so Laming disguised the device in a pair of binoculars, which his brother then "fired" from the grandstand rail, less than 140 meters from the finish line. With the push of a button on the binoculars, the miniature twenty-two-watt amplifier and loudspeakers apparently sent an ultrasonic blast directly into Ile de Chypre's ear, forcing him to swerve and throw Starkey off his back. According to the Associated Press at the time, prosecutors said "the ultrasonic device was the centerpiece of a plan in which narcotics profits would be laundered at tracks through bets guaranteed by use of the gun."

The race world was then further stunned (if you'll pardon the pun) to hear that the jockey at the center of the incident, Greville Starkey, was to appear in court. He was not part of the ultrasonic conspiracy, but he did agree to take part in tests to see whether the gun actually worked. Using three of his daughters' ponies—High Flier, Minstrel, and Dandino—as high-frequency subjects, Starkey rode them around his stud farm near Newmarket. He considered his daughters' ponies to be "bomb-proof," unflappable horses who were safe for young riders. The *Herald of Scotland* reported that Starkey said, "I got on the horse and went to canter him past the two gentlemen with the gun at a very leisurely pace, slow. All of a sudden he just took off. He took me two or three times around the paddock out of control." Starkey told the court that while riding toward two photographers, "I screamed out: 'I am out of control, get out of the swear word way.'" He also told Jonathan Goldberg, Laming's counsel, that "when we went through the tests yesterday it worked on two of my daughters' ponies very much the same as it did on Ile de Chypre."

Researchers suggest that the frequency range of horses' hearing is roughly between 55 Hz and 33 kHz. In the ultrasonic spectrum, there are around nine black and white Infinite Piano notes available to such a stun gun, between the upper extremities of human hearing (20 kHz) and horse hearing range (33 kHz). It therefore seems very plausible that Laming's binoculars/stun gun did nobble Ile de Chypre. The court, however, concluded that Laming's defense didn't make sense and that he couldn't really explain how his ultrasonic binoculars were related to the drug gangs he had connections with. Laming, along with his fellow defendants, was convicted with a substantial sentence. Whether the gun was ever used at Ascot remains an ultrasonic mystery, but there would have been a group of people present at the race who could have helped the police with their inquiries—teenagers and younger children.

Due to the presbycusis we all eventually suffer from, Laming actually had at his disposal many more than nine Infinite Piano notes in his stun gun range. The process of presbycusis begins in our early twenties and can lead to a serious hearing loss problem for many older people. It is a natural condition, due to the wear and tear on the hair cells in our cochleas. As the cells age and degenerate, our hearing ranges take a similar downward slide. Even by our midtwenties, most of us are already unable to hear anything above 18 kHz. Given the average age of a racehorse crowd, they could potentially have had an audible upper frequency limit of as low as 13 kHz. This would have given Laming a stun gun frequency range not of nine semitones but around sixteen. Though the gray-tops at the grandstand rail at Ascot wouldn't have heard any frequency between 13 and 20 kHz, young children and teenagers would have quickly stuck their fingers in their ears, proving beyond reasonable doubt that the gun was used at the King George V stakes in 1988.

The Antisocial Mosquito

The ability of young children and teenagers to hear the widest of human frequency ranges is both a blessing and a curse, an aural superpower and an Achilles' heel. When a Welsh teenager told her father how intimidated she was by a similar-aged gang hanging around the local shops, Dad—a former British Aerospace engineer—got to thinking. Instead of Howard Stapleton accompanying his daughter to the corner store and/or confronting the gang, he focused his paternal love through a frequency. Stapleton's childhood memory of an annoying high-pitched noise emanating from a local factory triggered his inventive streak, and he set about exploiting the teenagers' ability to hear very high frequencies. It wasn't a horse he wanted to jink—it was the local gang who were intimidating his daughter.

It was a genius idea. At around 17 kHz, his discreetly placed persistently squeaky "Mosquito" device could bug the hell out of youngsters, but anyone in their twenties or above would be completely oblivious to this very irritating high frequency sound. Subjected to a note almost exactly six octaves above the haunting, youth-defining "hello" that Kurt Cobain wails in "Smells Like Teen Spirit," those same Nirvana fans were soon scurrying from the sidewalk outside their local corner shop.

After some initial tests in 2005, a shop in Barry, south Wales, became the world's frontline in the battle against loitering "antisocial" teenage gangs. Who would have predicted that an aural Achilles' heel, possessed only by those who had yet to reach the age of twenty, could rid the world of menacing street-corner gatherings? Howard Stapleton's "Mosquito" invention was, almost overnight, a global media sensation. This sonic weapon augured a teenage-gang-free nirvana without any impact on more senior members of the population, most of whom already suffered from presbycusis. The brilliance of the invention was its simplicity. As the frequencies used are within "normal" hearing range, the tone generator and speaker were not specialized pieces of a kit. A 2005 BBC report focusing on one of the first installations at the Spar store in Barry explained how "the box, placed on the outside wall of the store, emits an 80-decibel pulsing frequency between certain hours." Robert Gough, who ran the store, said, "We just put it up and said nothing to the teenagers, then they came in to complain. They were literally begging me to turn it off. I told the kids it was to keep the birds away because of the bird flu epidemic. They waited for their friends and left covering their ears."

Even though verbal youth bashing was in vogue in the UK in the mid-2000s, Stapleton and his business soon felt a backlash... indeed, two of them. The first was from human rights activists and children themselves. Many felt that the targeting of a specific group of citizens who had done no more than possibly look intimidating was a breach of the 1998 Human Rights Act. In 2009, leading human rights organization Amnesty International produced an educational resource pack titled "Right Here, Right Now—Teaching Citizenship through Human Rights," which used this exact issue as a case study for classroom discussion. Whose rights are being violated: anyone under the age of twenty-five who happened to stop outside a shop, a shopkeeper whose business was affected by shoplifting and antisocial behavior, or the local person who was too intimidated to visit the shop in the evening? The 17 kHz of the Mosquito was such an incendiary frequency that it sparked widespread debate; a UK-wide campaign called "Buzz Off" was led by the children's commissioner for England, Sir Al Aynsley-Green, who said, "The use of measures such as these are simply demonizing children and young people." It was also discussed in the Houses of Parliament and the Council of Europe.

But the second backlash to the Mosquito was perhaps more of a classic (and ingenious) case of teenage rebellion. If 17 kHz was audible to young people but not to old fogies, then why not turn the tables? The adults' strength in their immunity to the Mosquito tone was transformed into a weakness by some very cunning school pupils, who realized that a 17 kHz ringtone could tinkle away completely undetected by teachers but was perfectly audible to kids who were supposed to have their phones turned off in the classroom. Text alerts could be received at any time during the school day, and much to the mirth of the pupils, collections of ringing phones could be sounded simultaneously, especially during drearily delivered physics lessons about frequency.

Various challenges were mounted to the implementation

of Mosquitos, such as the claim that they were a noise nuisance under the UK's 1990 Environmental Protection Act. They were also deemed to be inefficient in dealing with youth antisocial behavior as they didn't stop it but merely pushed the problem farther down the road. The UN Convention on the Rights of the Child said that the Mosquito's production of sound and implementation could be considered a form of corporal punishment. Shami Chakrabarti, director of the civil rights group Liberty, said in 2008, "What type of society uses a low-level sonic weapon on its children? Imagine the outcry if a device was introduced that caused blanket discomfort to people of one race or gender, rather than to our kids. The Mosquito has no place in a country that values its children and seeks to instill them with dignity and respect." The use of the phrase "low-level sonic weapon" by Chakrabarti was striking and undoubtedly deliberate (though perhaps in terms of frequency, "high level" would have been more accurate). But weaponized sound is nothing new.

Jericho's Trumpets

One of the most famous early examples of aural aggression can be found in the Bible. The Old Testament's Book of Joshua tells the story of the Israelites' conquest of Canaan. In the preceding books, the Israelites adopted new teachings and laws (the Covenant) following their emancipation from Egyptian slavery by Moses. On entering Canaan, Joshua and the Israelites' first target was the city of Jericho. The city was under lockdown, so Joshua needed a plan to gain access. According to the King James version of the Bible, God said, "And ye shall compass the city, all ye men of war, and go round about the city once. Thus shalt thou do six days. And seven priests shall bear before the ark seven trumpets of rams' horns." Joshua was instructed to circle Jericho, carrying and displaying the Ark of the Covenant (a gilded box that contained the two tablets of stone on which were written Moses's Ten Commandments), preceded by priests blowing rams' horns. It continues, "and the seventh day ye shall compass the city seven times, and the priests shall blow with the trumpets. And it shall come to pass, that when they make a long blast with the ram's horn, and when ye hear the sound of the trumpet, all the people shall shout with a great shout; and the wall of the city shall fall down flat." The ram's horn, called a shofar, undoubtedly becomes a weapon of war in this account and appears frequently in the Hebrew Bible and rabbinic literature.

Nowadays, the instrument continues to play an important part in Jewish religious life and is blown during High Holidays in the Jewish calendar, particularly around Rosh Hashanah (Jewish New Year) and at the end of Yom Kippur (the Day of Atonement). Though the shofar is made of a ram's horn, it goes through a surprising level of manufacturing processes to make the final product. Originally attached to the animal through a connecting bone, the larger end of the horn reveals a hole when removed from the dead ram. The small end of the horn is closed, so a hole is drilled through to the cavity at the larger end—it is also heat-straightened and polished. The sound is produced from the standing wave of the column of air, excited by the vibrating of the shofar player's lips, just as on any brass instrument. The shofar's frequency range is similar to that of a trumpet—roughly between 175 Hz and 930 Hz, from three notes lower than the first note of "The Last Post" to the high blast the trumpet climbs to at the end of Glenn Miller's "In the Mood."

In this range, even with a thousand shofars and a gigantic roar from the assembled Israelites, there would have been very little chance that the walls of Jericho would have come tumbling down (unless they were structurally unsound in the first instance). Most current scholarly research suggests that the Book of Joshua is probably not a factual account of historical events. But the idea of weaponized sound is repeated in the next book of the Old Testament, the Book of Judges. In it, the leader of the Israelites, Gideon, attacks a substantial Midianite camp with only three hundred men. He issues each of his men with a shofar and instructs them to blow on a given signal. The three hundred encircle the Midianite camp and, following Gideon's orders, blow their horns in unison. "Every man stood in his place all around the camp, and all the men in camp ran; they cried out and fled. When they blew three hundred trumpets, the Lord set every man's sword against his fellow and against all the army." In other words, the Midianites were terrified by the aggressive and nerve-racking racket of three hundred shofars blown in anger. Though the results of all this Biblical huffing and puffing were probably more effective on the Midianites' mental state than on the mortar of Jericho's walls, one cannot but question whether any enemy worth their salt would be scared by a mere noise.

In 1937, the Junkers Ju 87 entered combat in the Spanish Civil War, serving as a major new offensive weapon in the Nazi's Luftwaffe. By 1939, this dive-bombing plane—the Sturzkampfflugzeug—was creating terror over Poland in the early days of the Blitzkrieg. The Stuka dive bomber, as it is better known, had a sonic weapon that hurt no one but scared the living daylights out of everyone on the ground. The Stuka was fitted with "Jericho trumpets," two sirens mounted on its wings' leading edges, which terrorized ground troops and civilians with a pitch-rising crescendo of impending calamity. When the plane entered a steep dive, a propeller fitted to the front of the siren channeled air through the device. As the airflow increased, so did the pitch and volume of the siren. As with Gideon's shofars, the sound of these Nazi Jericho trumpets had a devastating effect on the morale of the enemy. It did not take long for the ground defenders to psychologically connect the cold-blooded wail of the Stuka with the threat of being blown to smithereens. The frequency range of the Stuka's siren was remarkably close to that of the shofar, but it was the unrelenting, unbroken rise of the pitch that gave it such a bloodcurdling, harrowing, almost human wail.

Wail Song

It is interesting to note that this wail (smearing through defined musical pitches) found a place in music as well as war during the twentieth century. Perhaps the most famous musical smear, which many claim ushered in the acceptance of jazz music as an art form, is the opening bar of Gershwin's *Rhapsody in Blue*. Following a low trill (that's a fast alternation of two adjacent notes), a solo clarinet rises through a scale until eventually it breaks out into a wailing slide across its final seven notes, dispensing with individual pitches to create the famous glissando up to its high B♭. This arguably signified a revolutionary moment in twentieth-century music, but was far from a blinding flash of inspirational genius.

The wailing glissando wasn't Gershwin's idea; he wrote out a rather boring scale for the solo clarinet to play. And it wasn't the idea of Gershwin's orchestrator, Ferde Grofé, who also faithfully copied out Gershwin's original, rather conservative opening. During the long rehearsals for the first performance with Paul Whiteman's band, clarinetist Ross Gorman decided to lighten the mood with a musical prank. He stretched and smeared the last notes of the scale into a slide, making the clarinet wail with a human vocal quality. According to Grofé, who was present at the rehearsals, Gershwin loved it. That moment of musical high jinks played a major part in the colossal success of *Rhapsody in Blue*.

In a world of very defined frequencies, musical sound that lies outside these twelve frequency straitjackets has an undoubted impact on the listener (Jim Reekes's detuned Mac start-up chime mentioned in chapter B♭ is a good example). Another composition that explores the nether regions of pitch—stylistically a world away from Gershwin's crossover masterpiece—is Krzysztof Penderecki's 1961 piece *Threnody for the Victims of Hiroshima*. It highlights how working outside established musical pitch can have a profound effect on human emotion.

The musical score is, at first glance, a strange and unnerving sight. On page one, there are no staves of five lines and four spaces, no key signature with sharps and flats, no time signature defining four beats in a bar or such, not even any notes, just a variety of symbols for the players to interpret. But it is this lack of defined pitch that gives the work its human intensity. Around one minute, forty-five seconds into the threnody, the unpitched chaos of the previous bar stops, and the cellos play a single note. They then begin to separate out (both upward and downward) with tiny microtonal slides, opening up a bone-chilling cluster of undefined pitches before they move back to their original note. The result is bloodcurdling, undoubtedly partly due to the slides that the cello players make. For Western ears, conditioned by hundreds of years of tonal definition, such smearing of pitches is unsettling and profound. Though Penderecki's *Threnody* is in no way a literal representation, at this moment in the piece, it is easy to imagine the sound of Enola Gay's engines as it approached its fateful target.

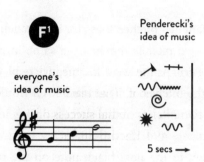

The Sound of Sirens

A similarly terrifying wailing sound of the Second World War became synonymous with the Blitz, the German bombing campaign waged against the UK between 1940 and 1941. This wail was not the sound of attack, though, but of defense: an aural signal to the civilian population to protect themselves from aerial bombardment. During the First World War, the authorities hesitated about introducing air-raid sirens. They were convinced that far from civilians running for cover at the sounding of the alarm, the sirens would encourage thousands to come out into the streets and watch the approaching zeppelins, making them perfect targets for the "baby-killers," as the British dubbed the airships. But in the lead up to the Second World War, the speed and ferocity of aerial attacks—as seen at places such as Guernica and Barcelona—meant that the implementation of air-raid sirens was essential to give clear and timely warnings to civilians, to keep bombing casualties to a minimum.

The distinctive and arresting sound of the sirens shares many similarities with the Jericho trumpet of the Stuka and the opening of Gershwin's *Rhapsody in Blue*. The pitch ascends through a continuous slide, passing through all frequencies within its range. However, what gives the air-raid siren a particularly striking and unique tone is not a single sliding pitch but two simultaneous pitches that rise in parallel. The distance between these two pitches remains the same in musical terms, a minor third. Perhaps the most inappropriate piece of music to help illustrate the pitches of an air-raid siren would be Brahms's "Lullaby." The two pitches of this sleep-inducing idyll are B_4 and D_5 at approximately 494 Hz and 587 Hz respectively. Only one single step down to the next black note on the piano and one arrives at the notes of the siren, B♭ and D♭ (around 466 Hz and 554 Hz). What is intriguing is that the two notes are a perfectly decent blend, a combination that has been used to great effect for centuries. It is not the pitches that create a sense of fear and foreboding but the rising, wailing slide the sirens create as they ascend toward these notes. Two different signals were sounded for the start and end of a potential bombing raid. To begin the alert, the siren would wail through a continuous rise (up to B♭ and D♭) and fall, while the all clear would rise to these notes and remain there for a longer period of time. But what was the significance of these two frequencies?

Such sirens have a relatively simple construction. Air is sucked into the machine and chopped up by rotating fan blades. The chopping of this air creates pulses of air pressure, individual waves that become a discernible pitch as the rotational speed is increased. As there are two separate fans on each machine, they each produce different notes. One fan has ten blades, the other twelve, and they both spin to a maximum of around forty-six revolutions per second. Multiplying forty-six by the number of blades that chop up the air reveals the pitches of each siren—for the ten-blade fan, it's 460 chops per second (460 Hz), the twelve-blade fan around 552 Hz. There is an intriguing additional aspect to the siren that I highlighted in chapter D♭. The difference between 460 Hz and 552 Hz is 92 Hz, and that frequency is just around

Gb_2. As the composer and violinist Giuseppe Tartini discovered, combining two notes together can often produce a "difference tone" or "undertone." What is particularly remarkable about this one—the Gb_2 at 92.5 Hz—is that the addition of this note to the siren's pitches at full revs makes a perfect Gb major chord (which can also be called $F\sharp$ major). When sounding the all clear, the sirens all sounded a celebratory Gb major/$F\sharp$ major chord: the first chord of the chorus of the Beatles' "Yellow Submarine."

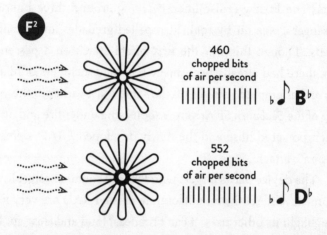

Though air-raid sirens were—and still are—a noncombative form of defense, not all defensive sonic weapons are quite as benign. On the morning of October 12, 2000, the U.S. Navy destroyer USS *Cole* docked in the port of Aden in Yemen for a short refueling stop. Within a couple of hours, a small craft approached the port side of the ship, carrying two al-Qaeda suicide bombers and over four hundred pounds of plastic explosives. At 11:18 local time, an explosion ripped a hole twelve meters by eighteen meters in the hull of the *Cole*. The attack killed seventeen crew and injured thirty-nine. The rules of engagement at the time forbade crew from firing at any possible threats unless they were fired on first, even after the initial explosion. In response to the *Cole* attack, a "nonlethal" defensive weapon was developed and deployed by the military and used by international commercial shipping and cruise liners that plied vulnerable waters.

In 2005, this weapon—named the Long-Range Acoustic Device (LRAD)—was sounded in anger to deter a terrifying pirate attack off the coast of Somalia. Very early on the morning of November 5, the *Seabourn Spirit* cruise ship was 160 kilometers off the Somali coast when it was approached by two speedboats. The luxury cruise liner, carrying around three hundred passengers, was hit by rocket-propelled grenades and a hail of bullets. Though this was the first pirate attack on a passenger ship, there had already been many recorded incidents of hijackings and attempted seizures off the same coast earlier in the year. Two of the *Seabourn Spirit*'s crew soon came under fire and, using a high-powered hose and the newly developed LRAD, deterred the pirate attack.

The device blasts targets with focused sound at ear-splitting volume levels. Yet the manufacturers of the LRAD are very keen to highlight its other uses. It can broadcast loud and clear spoken messages up to three kilometers away, undoubtedly invaluable in life-threatening situations. An ability to communicate effectively over such large distances can certainly play a key role in disaster zone and search and rescue missions. But the LRAD is probably more known for its controversial "alarm tones."

Protestors at the 2009 G20 summit in Pittsburgh caught an earful of one such alarm in what is thought to be the first police department use of the LRAD at a high-profile event. Online footage shows how everyone almost instantaneously covers their ears as the first blast is sounded. My own frequency analysis backs up other research that suggests that the tone generated sweeps between approximately 2,300 Hz and 3,200 Hz. This is in a similar pitch range to the final few whipped scales played by the piccolo at the end of Beethoven's *Egmont* Overture. The highest note in each of these runs is an F at almost 2,800 Hz, which can prove to be a most uncomfortable few bars for other orchestral musicians in close proximity. And this is precisely why some opponents of the LRAD raise concerns about the connection between the alarm tone frequencies and the sensitivity of the human ear.

In 1933, Harvey Fletcher and Wilden A. Munson published a paper titled "Loudness, Its Definition, Measurement and Calculation" in the *Journal of the Acoustical Society of America*. Harvey Fletcher was a particularly notable scientist who worked with the great conductor Leopold Stokowski to develop the first stereophonic recordings, the first live stereo sound transmission, and the first vinyl recording. He also invented an electronic hearing aid and an artificial larynx. He and Munson discovered that sounds that are equally loud but at different frequencies are perceived by our brains as having different levels of volume. Their graphic curves have been further refined by others, but the initial principles remain the same. As we climb from 20 Hz to just above 1 kHz, we perceive such frequencies as increasing in volume. From around 2 to 5 kHz (just above the highest note of a standard piano), our ears are at their most sensitive, and we hear these frequencies as the loudest. After 5 kHz, our aural sensitivity drops once more, and we perceive higher frequencies as being quieter than the midrange ones. One of the reasons that our ears are most sensitive to the 2–5 kHz range is due to the physical properties of our ear canals. Our auditory canals are closed tubes, so they resonate just like any other similar structure. With an average length of around 2.25–2.5 centimeters, our ear canals have fundamental frequencies around 3.8–3.9 kHz (two notes from the top C on a piano). Even though we might play top and bottom Cs on a piano at exactly the same volume, our brains will always perceive the top note as being louder.

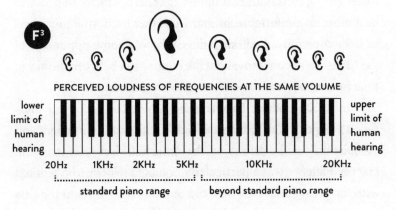

Fletcher and Munson loudness curves

And it is for this reason that critics of the LRAD system have been so vocal. Of all the frequencies that could be utilized as an alarm tone, they question why those in the most sensitive region of human hearing have been chosen. The National Center for Biotechnology Information states that a sound pressure level of 110 dB is the threshold of discomfort, and 130 dB is the threshold of pain. It also warns that sounds louder than 130 dB can cause acute hearing loss. Reports suggest that an LRAD can emit a beam of sound between 137 dB and 162 dB; it is worth remembering that every 10 dB increase is perceived to have a doubling of loudness. But perhaps LRADs should not be our major concern when it comes to our aural well-being. The World Health Organization suggests that "some 1.1 billion teenagers and young adults are at risk of hearing loss due to the unsafe use of personal audio devices." Are we deafening our own kids with excessively loud music from giant speakers or earphones stuffed deep into their ear canals?

Doppler's Shift...on a Train

An important train journey in 1845 will help explain the invention of our next sonic deterrent. On June 3 and 5, a steam locomotive on the line between Utrecht and Maarssen in the Netherlands helped to ground a celestial theory. Christophorus Henricus Diedericus Buys Ballot was a Dutch chemist and meteorologist whose trackside sound experiments would have seemed a tad quirky to nineteenth-century Dutch train spotters.

The apparatus for the tests did not include this chemist's usual test tubes, chemicals, or even a meteorologist's wind sock. For this enterprise, Buys Ballot gathered a gaggle of brass-playing musicians, a locomotive train with a single flatbed truck, a bell, a station, and a number of listeners with keen musical ears. Though this talented scientist went on to achieve fame through his Buys Ballot meteorological law (which validated the theory that the wind travels around a low-pressure zone in an anti-clockwise rotation in the Northern Hemisphere and clockwise in the Southern Hemisphere), Buys Ballot was trying to prove astrophysical principles with a steam train and a bunch of local trumpeters.

He wanted to extend Austrian physicist Christian Doppler's astronomical theory of 1842. In Doppler's treatise "Über das far-bige Licht der Doppelsterne und einiger anderer Gestirne des Himmels" ("On the Colored Light of the Binary Refracted Stars and Other Celestial Bodies"), he proposed, in effect, that stars were like ducks in a pond after you have shouted "Boo!" at them. (It is worth noting at this point that any reference to ducks is mine and not Doppler's—he used a shipping analogy.) If I were feeding bread to ducks in a pond and then scared them, they would all swim away from the bank in different directions. Similarly, stars are also moving away from us due to a louder "Boo!" called the big bang, which has resulted in the ongoing expansion of our universe. When a duck swims, it creates waves; the waves in front of it bunch up (it is catching up with the waves it has created), and the waves behind it spread out (it is leaving behind waves it has generated). This means that the frequency of waves in front of the duck is higher than the ones behind it. Doppler's interest was more light waves than pond ripples, but the principle is the same. And in terms of visible light, higher frequency light waves are in the bluer part of the spectrum, while lower frequencies are in the reds. Doppler was, in essence, observing the back end of ducks as they traveled away from him, their bottoms emitting red light, not blue.

Doppler suggested that sound waves would act similarly if the speed of an object emitting such waves traveled fast enough, and Buys Ballot couldn't resist trying it out with a train. It might seem a little curious to the twenty-first century layperson that the discovery of the Doppler effect came initially from astronomical observations and not from the Doppler effect we hear daily from our emergency service vehicles. But it is worth remembering that the vital ingredient of this natural phenomenon was absent from early nineteenth-century life—speed. Opportunities for people to witness the Doppler effect were very rare, and it was only the invention of the speedy locomotive train that allowed the acoustic phenomenon to be experienced and tested.

Buys Ballot wrote an account of his experiments in the *Algemeen muzikaal tijdschrift van Nederland*, published on August 15, 1845. He spends quite a large portion of the article trying to explain how all semitones and tones aren't equal and how the ratios for all of these are constantly changing. As discussed in chapter B♭, pitch was a moveable feast, and an experiment relying on nothing better than keen musical ears and potentially out-of-tune brass instruments was perhaps not the most scientific of methods to prove a celestial thesis. All the same, Buys Ballot's results confirmed Doppler's theory that the frequency of a sound wave increases if the object making it is approaching you and decreases if it is moving away from you. Buys Ballot, with a slightly tiresome tone, wrote "perhaps it is still necessary to point out to the reader that, during the approach, the note was always heard to be higher and also always higher than the actual note that was blown, while the opposite happened during the departure." Conductor Charles Hazlewood recreated the experiment in an entertaining 2017 BBC radio documentary, "The Doppler Effect with Charles Hazlewood"—the Gs blown by the brass players rose toward a G♯ as the train approached and flattened to an F♯ almost instantly after the train rushed past. Writing in a serendipitous final paragraph, Buys Ballot is honest in his acknowledgement that "it is impossible to know what purpose the findings in this study could serve, but neither is it known how a newly-found truth can sometimes unexpectedly find its application."

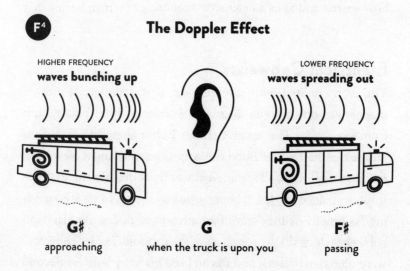

F⁴   **The Doppler Effect**

HIGHER FREQUENCY   LOWER FREQUENCY
waves bunching up   **waves spreading out**

G♯   **G**   **F♯**
approaching   when the truck is upon you   passing

Perhaps his only mistake here is the omission of a plural at the very end. His head would undoubtedly be spinning (in an anticlockwise motion if he was in the Northern Hemisphere) at the dizzying array of applications the Doppler effect has enabled, from the police speed gun to the echocardiogram, which could have generated images of Buys Ballot's excited heart rate. But perhaps the application that would have given him the most pleasure would have been meteorologists' ability to "see" weather by bouncing frequencies of between 2 and 4 gigahetz into the atmosphere and measuring the return waves, some of which may have a higher or lower frequency due to the rain, clouds, or even air particles moving toward or away from the weather radar station...every cloud has a Doppler lining. It is understandable that neither Doppler nor Buys Ballot could have foreseen the amazing range of applications that an understanding of this phenomenon would lead to, be that anything from gynecological monitoring to deterring burglars. And when it comes to keeping such miscreants out of our homes, the Doppler effect plays a major part in how we use sound as a defensive weapon.

Electrical Cobwebs

The man credited with inventing the first commercial motion sensor was one Samuel Bagno. A wonderful 1950 photograph from the *Denver Post* archive shows Bagno alongside Mrs. Sara Jackson of the Denver Burglar Alarm Company and a New York city insurance man, all smiling awkwardly at the camera, the contraption placed on the table in front of them. They are standing "as quietly as they can in an attempt to dodge the electrical cobwebs which fill the room," says the caption. Those "cobwebs" were ultrasonic waves, and Bagno used his knowledge of Second

World War radar systems, ultrasonic waves, and the Doppler effect to develop a commercial and domestic appliance.

Bagno's motion sensor generated an ultrasonic frequency similar to James Laming's binoculars (the ones that reputedly nobbled the racehorse Ile de Chypre). When the "sound" bounced back to his machine from another surface, it remained at the same frequency. However, if a person disturbed the wave by moving toward the sensor, the Doppler effect would mean that the reflected wave peaks would be squashed together a little (think ducks), raising the frequency of the sound very slightly above the initial ultrasonic note. An alarm would then be triggered. Conversely, if the intruder were moving away from the emitted ultrasonic wave, its frequency on the reflected wave would be lower, and once again the sensor would trigger an alarm. In a most informative article in the journal *American Scientist*, Eugene L. Fuss calculated that "an intruder walking at one step a second generates a Doppler shift of about 65 Hz. If he can move very slowly indeed, he may never generate a shift of more than 25 Hz, but if he makes the slightest uncontrolled movement he may be detected." Of course, air can move on its own, perhaps due to heaters or drafts, and Fuss noted that "relatively severe turbulence can produce Doppler shifts of 25 Hz, and therefore most systems use electronic filters to ignore any changes of less than this amount." Soon, however, the limitations of ultrasonic motion sensors were overcome by the use of microwaves and infrared, a massive jump in terms of frequency. To play an A on the Infinite Piano in the infrared zone, one would have to travel around 4.5 meters to the right of middle C to sound the frequency of 59,055,800,320 Hz.

Deterring criminals is all well and good, but what if you want to catch them in the act and collect a piece of evidence? Another neat ultrasonic application, like that of Bagno's, was introduced to the world in 1978, simplifying and revolutionizing the way we capture images.

The Revolution of Autofocus

Six years earlier, the camera company Polaroid began selling an innovative new folding camera that automatically processed and ejected a photograph snapped only moments before. Naturally, the SX-70 was a big hit. No longer did you have to dabble with nasty chemicals in a darkened room or wait for many days to have your personal memories (and possibly intimate pictures) developed and leered over by a pimply kid at the local drugstore. The SX-70 became the camera of choice for a generation for whom instant gratification was a must. Not only was it the camera to be seen in front of, it was also the camera to be seen *behind*. In an awkwardly posed celeb picture of 1973, Andy Warhol was snapped by professional photographer Garry Winogrand at novelist Norman Mailer's fiftieth birthday bash in New York City. What is striking about the image is that this other icon of the era is also featured in the shot. Held close to Warhol's chest is the unmistakable SX-70. The fold-up, darkroom-in-your-hand, amateur photographer's dream was everywhere. But its one drawback was that it was very difficult to focus, especially in low light; using the SX-70 to snap a sharply focused shot of our burglar as Bagno's motion sensor sounded an alarm would be nigh on impossible.

In 1978, Polaroid took inspiration from bats, dolphins, submarines, and Bagno himself, solving the focus problem with the introduction of the SX-70 Sonar OneStep Land Camera. This camera had all the features of its predecessor, retaining its iconic shape and continuing to allow complete novices to load prepared film cassettes, avoiding all that faffing with pesky coiled film. But from this moment on, complete novices didn't even need the photographic skills to focus their own shots. For some, 1978 was the beginning of the democratization of photography, while for others, it was the start of the terminal decline in photographic technical skill that has led to latest estimates suggesting over a trillion photographs are taken annually (most of which are judged to be technically "crap"). The SX-70 Sonar OneStep's TV commercial stated very proudly, "When I press this button, a new era in photography begins." Over a 1970s burbling orchestral soundtrack reminiscent of *Star Trek*, the narrator continued, "This camera sends out inaudible soundwaves that bounce off the subject and return in a split second. The lens automatically rotates to perfect focus." He then summed up with a rather laughable promise (by twenty-first-century standards) that "you can get a precisely focused picture every time in minutes, at the touch of one button." One suspects there would be riots on the streets if any present cell phone company suggested that customers could have focused pictures on their phones *in minutes*.

Polaroid's inclusion of the word *Sonar* in its name gave the camera a hint of futurism, Cold War stealth, and scientific glamour. Sonar had been utilized throughout both world wars, mainly in nautical areas of battle. As mentioned in chapter D, sonar was invaluable to the military during the twentieth century and was especially vital for monitoring enemy submarine activity during the Cold War. There are two types of sonar: passive sonar, which involves very quiet listening for the sound of ships (something Leonardo da Vinci had reported on following his experiments sticking tubes in Venice's Grand Canal in the 1490s), and active sonar, whereby sound is transmitted and its reflection evaluated, as in Bagno's motion sensor.

A fantastic technical article of 2002 explores the history and physics of the SX-70 Sonar OneStep Land Camera, giving us a comprehensive analysis of the camera's workings. In it, authors Dan MacIsaac and Ari Hämäläinen wax lyrical about the camera's ingenuity, explaining that it has "a single transducer that acts as both the ultrasonic signal source (loudspeaker) and detector (microphone)." The transducer is the circular assembly of gold-coated foil and an aluminum plate positioned on the top of the camera. "When a high-voltage signal (a 49.4 kHz square wave at 300 to 400V) is applied to the transducer assembly, the foil is driven back and forth by electrostatic forces creating ultrasonic pressure waves in the air." This is the beam of sound that bombards the smiling subject with an ultrasonic frequency, at 49,400 Hz; this equates to the G that Sting sings at the start of "Roxanne," seven octaves higher (on our Infinite Piano, this is only just over a meter above Sting's note but way beyond human hearing). Because of its ultrasonic frequency, a quick family photo of toothless Granny on her birthday would result in a lovely, naturally posed touching portrait—Granny would be completely unaware of the sound beam being blasted at her. However, a quick photo of the family's pet rat Pinky on its birthday would probably reveal a very different facial expression, as 49,400 Hz is well within the rodent's hearing range. Pinky's birthday photo would feature a rat with a quizzical expression, asking why the photographer had gained the superhuman quality of being able to sing seven octaves above Sting.

The transducer assembly on the camera rapidly changes from a sonic emitter to a sonic receiver in only a couple of milliseconds. The article explains that after the sound has been transmitted, "the transmitter circuitry shuts down, and receiving circuitry is turned on... The interface or calculator that is running the transducer starts an elapsed-time clock." As the sound is partly reflected by Pinky the rat's head (some of it is absorbed by the rat as well), the elapsed-time clock is stopped when the reflection arrives back at the camera's transducer assembly, which has now become an ultrasonic ear. The camera then calculates the elapsed time and the speed of sound and, hey presto, determines how far away Pinky the rat is. But its work remains incomplete. As Pinky the rat holds his quizzical expression for a couple more milliseconds, the camera alters the lens's focal length to sharpen Pinky's image.

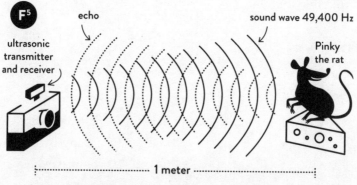

Camera calculates distance to Pinky by how late the echo is

At this point, the actual picture taking begins, and Pinky's birthday pose is captured for posterity...except for one possible problematic scenario. If Pinky was kept in an aquarium (yes, apparently some rats are, presumably without the water), then the 49,400 Hz sound would bounce off the glass. The camera would focus on the surface of the glass and not on Pinky and his birthday cake farther back inside the aquarium.

This was one of the drawbacks of the SX-70 Sonar OneStep too—you couldn't autofocus through a window. So, as a sneaky cat burglar tries the lock on my patio door, triggering the Samuel

Bagno motion sensor I have installed, the SX-70 Sonar OneStep Land Camera I have positioned in the living room is set off, snapping a photograph of the intruder. Unfortunately, though, as the camera is inside the house, it focuses on the glass of the patio door, producing a blurred and therefore useless image of the cat burglar. However, if the miscreant was truly a cat burglar, they might have heard the ultrasonic sound in the first place.

SLEIGHT OF EAR

Frequencies that ain't what they seem

Some years ago, I was working as an arranger on an album that was packed with celebrities and musical luminaries. The project was bursting with wacky musical challenges across a dauntingly broad range of genres. It culminated in a most memorable day at Abbey Road Studios: I conducted an orchestra in Studio 1 in the morning; after lunch, I hung around in the control room of the famous (thanks mainly to the Beatles) Studio 2, observing some great sessions; and eventually I joined a few of the string recordings toward the end of the day's very hectic schedule.

About a week later, I got a call from the producer of the album, asking me about pitch correction software. On listening back to the vocal sessions from the day, his worst fears had been confirmed. The vocal takes of one particularly well-known celebrity—who was key to the project—were awfully out of tune. Following several torturous weeks of painstaking editing (the producer preferred the term "butchering"), the suspect vocal tracks gained some semblance of musical respectability, and the finished song was presented to said celebrity. The two sat in silence, listening intently to the recording. Undoubtedly the producer never wanted to hear this song again, but the celeb in question remained focused until the very last note. He then turned to the producer and said, "Well, that sounded much better than I remember it on the day. Perhaps I've got a better voice than I think I have. Let's make an album!"

Therein lies the difficulty most people have with pitch correction—it's a double-edged sword. Perhaps the best illustration of the pitch correction conundrum is the European Commission's Regulation EEC No. 1677/88, concerning the quality of cucumbers (yes, I really think it might be the best illustration). "Class I" cucumbers were to be, among other things, "well shaped and practically straight (maximum height of the arc: 10 mm per 10 cm of length of the cucumber)." We all scoffed at the ridiculous notion that an organization would attempt to regulate the uniformity of fruit and vegetables. A BBC News article of 2008 asked the question, "Will we eat wonky fruit and veg?" Of course we will, won't we? Everyone would agree that it is completely natural that vegetables and fruit grow to different sizes, with slightly different colorations, individual blemishes, different arcs of bendiness—that's nature. But here I am, a twenty-first-century studio musician, spending considerable amounts of time in darkened soundproofed rooms, using the latest technological gimmickry to straighten musical cucumbers, coloring all voices to similar hues, feeding some sounds to make them fatter than they actually are, and dehydrating others to thin them out.

I hang my head in shame on so many occasions, acutely aware that I am perpetuating the monstrous myth of musical perfection. But wait a doggone moment; do not think that you, the reader, are responsibility-free. Just as customers at supermarkets leave the wonky carrots and bent cucumbers at the bottom of the plastic boxes, so music listeners turn their noses up (or at least their ears) at the ugly stench of an out-of-tune vocal note. And this attitude is now so pervasive that I would argue it's having a detrimental effect on the creative process itself. I work with a lot of young musicians in studios, and virtually all of them see intonation (the ability to sing or play "perfectly" in tune) as of paramount importance. It is of no consequence to them whether their carrots or cucumbers taste any good: it's only about whether their product complies with the "rules." And this is because, for a lot of young performers, they only ever listen to straightened cucumbers!

How to Pitch a Revolution

One could argue that the rot started in 1996 when a fascinating mathematician, inventor, electrical engineer, geophysicist, entre-preneur, musician (I could go on) called Dr. Andy Hildebrand turned his attention to computer-led pitch correction. His career is an eye-watering series of seemingly unconnected innovative triumphs, from utilizing data on alfalfa weevils to improve the effectiveness of crop spraying to saving the Exxon oil company $500 million by fixing their faulty seismic instrumentation. In 1982, he founded a company called Landmark Graphics that spe-cialized in 3D graphic workstations for oil companies and others; it had outstanding expertise in the interpretation of seismic data. Hildebrand once said of this work, "I often describe it as stand-ing outside in a thunderstorm, listening to the thunder, and then computing the shape of the clouds." However, not satisfied with the success of his world-leading work in geophysics, he retired from Landmark to pursue a lifelong passion, enrolling on a music composition degree course (Hildebrand learned the flute from the age of twelve).

He was particularly interested in composition involving sampling synthesizers, though he rapidly realized that these machines were woefully inadequate. Once again, he turned to the extraordinary skill that ran through all his success to date, his talent in mathematics—very, very complex mathematics. At a National Association of Music Merchants (NAMM) conference in 1995, someone joked to Hildebrand that he should invent a box that could allow them to sing in tune. The conversation moved on, but some months later, Hildebrand set his mind to the task.

Many before had tried and failed due to the incredibly high level of mathematical computations necessary to analyze and correct a human voice in real time. In 1995, the challenge still seemed impossible; it would require a vast level of computational processing power. But the combination of Hildebrand's previous work with gigantic data sets in seismic analysis, his mathematical genius, and his background in music allowed him to find the solution using a process called autocorrelation. By the NAMM conference of 1996, producers were desperate to buy his new software product, Antares Auto-Tune, and one of the most controversial developments in modern music was born.

So how does Auto-Tune work? The principle, as ever, is relatively simple. At its core is the combination of the International Organization for Standardization's A at 440 Hz and equal temperament's fixed musical frequencies (for example, $B_4 = 493.88$ Hz, $C_5 = 523.25$ Hz, $C\sharp_5 = 554.37$ Hz, $D_5 = 587.33$ Hz, etc.). Auto-Tune gives each of these set frequencies a sort of magnetism. Let us imagine a scenario where the great Luciano Pavarotti sings his first E_4 in Verdi's aria "La donna è mobile" out of tune— inevitably he did on occasions, because he was a human (but it was still fantastic). The recording software analyzes Pavarotti's flat E almost instantaneously, noting it as wobbling around 321 Hz. The magnetic pull to an absolute E_4 at 329.63 Hz will be stronger than the nearest frequency of Eb_4 at 311.13 Hz. The only other factor Auto-Tune considers at this stage is how strong the force of the magnetic attraction is. The software can be set to attract the out-of-tune note to the "correct" frequency at a slow rate that sounds relatively "human" or an almost instantaneous correction within only a few milliseconds. At first, producers and artists used it as a subtle touch-up tool, speeding up the previously laborious task of cutting out the best bits of different takes of a singer's performances to stitch them together and make one "perfect" take. But in 1998, Auto-Tune was thrust from the dimly lit corner of the studio into the international limelight. Overnight, it became a *creative* tool rather than a remedial one, and it made virtually every musician stop in their tracks.

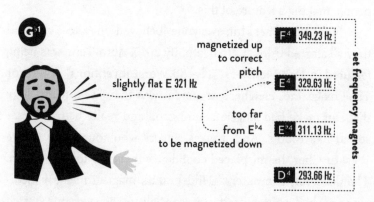

The transformation happened on Cher's "Believe," the best-selling single of 1998. Apart from the radical change of style that this single heralded for Cher, much of its impact and popularity undoubtedly came from *that* sound. After a very run-of-the-mill pop introduction, Cher starts her first verse with the words "No matter how hard I try, you keep pushing me aside," and then

BAM!—"and I *can't break through.*" It's very easy to forget the wow factor of those three robotic words. "How have they done that?" was *the* question being asked by commercial artists, musicians, producers, and members of the general public worldwide. In a 1999 interview, the producers of the track, Mark Taylor and Brian Rawling, fibbed about how they had created Cher's stepped, robotic vocal; they wanted to keep their trade secret. The whole point of Hildebrand's invention was to hide this surreptitious software in the shadows; this was the secret technological glitter in which to roll singers' turds. Hildebrand had given users of the software the freedom to choose the rate at which out-of-tune notes could be "magnetized" to the correct pitch. But he'd also left the most extreme setting of zero on the plug-in, where the software would instantaneously correct any pitch defects. I suspect that never in his wildest dreams did he foresee creative people making a feature of this.

Soon after Cher's hit, everyone "believed" in Auto-Tune, and they all started using it...to death. By 2010, Auto-Tune was being featured in Dan Fletcher's "The 50 Worst Inventions" article for *Time* magazine. Fletcher's argument was that "it's a technology that can make bad singers sound good and really bad singers... sound like robots. And it gives singers who sound like Kanye West or Cher the misplaced confidence that they too can croon. Thanks a lot, computers." Hildebrand's magical musical medicine now had a bitter aftertaste, especially when everyone started using it on live performances as well as studio ones. Hildebrand's counterargument—which is well placed in terms of the history of recorded music—was that in pre-Auto-Tune days, the same process (which many call "cheating") was followed, albeit in a much more time-consuming and laborious manner. Who would want to sing or record over a hundred takes of a song and chop it all up to find the best bits when they could simply push a few buttons over a couple of renditions?

A B or Not a B—That is No Longer the Question

In 2015, I made a film for the BBC about the controversy that continues to surround "pitch correction." (Auto-Tune is a brand name, something that is avoided on the BBC.) I recorded a backing track of Queen's "Don't Stop Me Now" and invited a random group of eight members of the public to a recording studio to see whether I could convincingly make them sound like "proper" singers. The eight ranged in vocal talent from "I can hold a tune" to "I've never sung before." (Genuinely, he claimed he'd never sung before. I believed him once he started.) After spending hours in postproduction ("butchering"), I managed to find enough workable snippets from each vocalist to create a half-decent vocal montage in terms of pitch accuracy (though one could hear the moments when the software was working in overdrive). There was no doubt that everyone was in tune, but their newfound accuracy added nothing to the overall musical satisfaction of the performances. However, I avoided confusing the main thrust of the short film, conveniently ignoring the newest generation of pitch correction software, the technology of which is truly astonishing.

In an artistic sense, Auto-Tune allowed humans to draw perfectly straight lines by tapping the pencil back on to the correct path when it strays, all in real time. Modern-day pitch correction software goes much, much further. It allows you to draw a line freehand and then review every micron of the pencil mark, editing it accordingly at a later moment. You can now tweak errors that might stray outside acceptable tolerances but still retain a modicum of human "wobbliness" throughout the rest. In such an instance, the observer would believe this to be a hand-drawn line because of its tiny imperfections; they would marvel at the creator's almost superhuman ability to draw so accurately. Pitch correction software means that you can, for example, raise the first twenty milliseconds of a note by three one-hundredths of a semitone and then flatten out the slightly suspect vibrato toward the end of that same note without affecting any other pitch around. Indeed, the software is now so advanced that you can tune an individual note within the recording of a guitar chord, even though it might contain five other notes. And for good measure, a performer's rhythmical inaccuracies can be airbrushed as well, as notes can be moved earlier or later along the timeline, stretched or squashed to fit, or even magnetized to the nearest beat.

The heart of the controversy behind modern pitch correction software is that it is impossible to know whether a recorded sound has been edited or not—is it real or is it merely a confection? If modern recordings are blemish-free but sound perfectly natural, young performers assume that these are "real" recordings of musicians with extraordinary musical superpowers. How incredibly talented these shiny stars of the musical firmament are, these superhumans who can grow straight cucumbers and draw perfect straight lines. Hearing my musical idols missing a word in a song or playing a note slightly out of tune is inspiring, a chance to see that even great artists still struggle. This struggle is what makes all art a human endeavor. Hildebrand's genius in applying advanced mathematics and frequency to a musical tool of revolutionary importance is undoubted. But as so often is the case in science, blessings can quickly become curses.

To defend himself, Hildebrand recently said, "I just build

a car, I don't drive it down the wrong side of the freeway!" He draws a parallel between math and music. "Pick a university, walk into their symphony orchestra rehearsal, ask everyone there to raise their hand if they're in engineering or mathematics, and half that orchestra will raise their hand. Why is that? I believe it's that music and mathematics have a common denominator...and that's symbolic abstraction"—in other words, the ability to represent, imagine, and manipulate the abstract.

The Eternal Sounds of Shepard and Risset

The presentation of statistical data in a form that can be understood by humans—symbolic abstraction—is something that American cognitive scientist Roger Shepard has pioneered. Like Andy Hildebrand, Roger Shepard is a relatively unknown name, but his impact on the wider world has touched most of us, albeit for Shepard through the medium of film music.

Shepard's brush with fame in this genre is most notable in the 2017 war epic *Dunkirk*, directed by Christopher Nolan. The score was composed by Hans Zimmer, who uses what is known as the Shepard tone to underpin a prevailing atmosphere of rising tension. In essence, the Shepard tone is an aural version of the barber's pole, the column of seemingly ever-rising red, white, and blue diagonal lines. Though we know that this is simply a spinning column that has no real vertical motion, we cannot persuade our brains of this—the incoming visual stimulus is interpreted as an infinite ascent of red, white, and blue lines. Similarly, major sections of the score of *Dunkirk* seem to rise eternally. But how did Shepard come up with the tone in the first place? What was a cognitive scientist doing dabbling with tension-inducing frequencies?

Shepard's interest in how organisms learn through generalization led him to try to map their responses. How *do* we and other animals learn? How do we gauge whether our own response to one situation will work in a later similar situation? If the scenario is not the same, why would our first response work in this new situation? Similarly, how do birds or frogs or dolphins learn? This is where Shepard developed a system for mapping and illustrating abstract possibilities in such a way that humans could interpret the data visually. Some of the most entertaining (though no less important) parts of Shepard's work have been his sketches, not dissimilar to those of the enigmatic Dutch artist M. C. Escher, famed for his mind-bending monochrome pictures of people walking up descending stairs (and down ascending stairs). Shepard's interest in the cognitive processes that the mind uses to mentally rotate, for example, a 3D shape, led him to explore how our brains process auditory information. The Shepard tone is a perfect "sleight of ear," an auditory conjuring trick involving a musical scale, each step of which contains a number of notes in octaves. Imagine three different As, each an octave apart, played simultaneously, followed by three Bs played in the same fashion, etc. As you move up the scales, the highest octave note in each step is incrementally faded, while the lowest octave of the three is raised in volume. By the time another A comes around, the blend of the three octave notes sounds exactly the same as the initial step. In this way, the scale seems to be infinitely rising, as our attention is drawn from the highest climbing notes to the lower ones.

The internet is awash with amateur film critics explaining the use of the Shepard tone in *Dunkirk*. However, the film does not contain the Shepard tone—it contains something called the Shepard-Risset glissando. Jean-Claude Risset, who died in 2016, was a French composer noted for his work with computers. As well as studying music, he earned a doctorate in sciences and worked at Shepard's stomping ground, Bell Labs in New Jersey. Risset built on Shepard's scale by creating a sliding version, one without steps, an escalator instead of stairs (much like the smeared wail of *Rhapsody in Blue* and the Second World War air-raid siren). The Shepard-Risset glissando is the eerie sound that is heard in *Dunkirk* and similarly in another of Christopher Nolan's films, *The Dark Knight*, during which the ever-rising engine tone of Batman's motorbike, the Batpod, creates an unnerving and menacing atmosphere.

Tchaikovsky's Musical Tricks

The notion of sleight of ear through such things as infinite scales is not particularly new. The nineteenth-century Russian composer Pyotr Ilyich Tchaikovsky is best known for works such as "Dance of the Sugar Plum Fairy" and the 1812 Overture. Such lollipops of the classical canon do not necessarily truly represent this most ingenious writer though. One of Tchaikovsky's trademarks is, in some ways, the precursor to the Shepard tone.

His sixth symphony, the *Pathétique*, is littered with dove-tailed "infinite" scales. Except for our Infinite Piano and perhaps a synthesizer, every instrument has a limited range of notes it can sound. The violin has quite a broad range, from G_3 at 196 Hz (the first note played by the solo violin in Max Bruch's Violin Concerto no. 1) to over four octaves higher, around A_7 at 3,520 Hz. Others, such as the trumpet, have a much more limited range, from F_3 at 175 Hz to C_6 at 1,047 Hz. (Though the likes of jazz trumpet virtuosi Maynard Ferguson, Wayne Bergeron, and Arturo Sandoval would most certainly disagree with this upper limit. In his track "Sandunga," Sandoval hits a stratospheric G_7 at 3,136 Hz.) To exploit wider instrumental ranges, Tchaikovsky often treated instruments as family sections rather than as individuals. For example, when combined, the string section has a whopping range from the double bass's E_1 at 41.2 Hz through to the highest of violin notes.

The *Pathétique* Symphony contains numerous examples of moments where Tchaikovsky begins scales on the lowest instrument of a particular orchestral section, then dovetails the entry of the next available instrument as the initial one starts to reach its upper limit. During the first movement, there is a beautiful imitative rising scale conversation between a flute and bassoon over an accompanying soft bed of bouncing strings. After four bars, Tchaikovsky starts a little counter melody, a scale in bass trombone and tuba. Four notes later, the tuba is dispensed with and the bass trombone takes over its role. At that same moment, he introduces another trombone. Within five further notes, he does a similar trick, this time losing the bass trombone completely and adding a trumpet low in its range to continue the higher octave of the scale. Tchaikovsky achieves a similar effect in the second movement (whose innovation in terms of its meter of five beats in a bar is often overlooked; he was twenty years ahead of Holst's "Mars" from *The Planets*), with a rising scale being passed from cello to viola to second violin to first violin, all in the space of four bars. In the third movement, a stirring march, Tchaikovsky pulls out all the stops for his rising scale, on this occasion dovetailing instruments across the orchestra as the music gears up for its final rendition of the main theme. It starts in bassoons and cellos and violas and quickly adds clarinets and oboes and violins, all entering at different times. The resulting aural effect for the listener is a long, rising anticipatory scale, relatively indiscernible in terms of its instrumentation but a most effective musical precursor to the climax of the movement.

But Tchaikovsky has yet more sleight of ear tricks up his sleeve. At the beginning of the final movement, he presents a theme that predates the research of another major cognitive scientist, one known for her work in aural illusions, Diana Deutsch. Born in London in 1938, Deutsch studied psychology, philosophy, and physiology at Oxford before moving to San Diego to continue her research. She increasingly specialized in how our brains process music and sound, looking particularly at "top-down processing," where our experiences, beliefs, and expectations shape our reactions to music. After setting up her research lab, Deutsch developed a paradigm in which two different sequences of tones were repeatedly presented to the listener via headphones, such that when the right ear received one sequence, the left ear received the other sequence. She found, to her surprise, that "striking illusions occurred under these conditions." What is startling, though, is that Tchaikovsky must have known about this same aural illusion seventy or eighty years earlier. To understand this aural trick, we must first consider the seating pattern of the nineteenth-century orchestra.

In the UK and many other parts of the world, orchestral string sections are placed with the first violins on the left (as we see them from the audience), second violins left of center, violas right of center, and cellos and basses on the right. However, many European and Russian orchestras place the first violins extreme left and the second violins extreme right, giving a perfect balance between violins if you were listening on headphones (not available to Tchaikovsky).

The final movement of the *Pathétique* begins with an angst-ridden, passionately yearning string melody... Well, it does and it doesn't. Imagine two people in front of you, one on the left, one on the right. When the left person is standing, the one on the right is sitting and vice versa—this is how Tchaikovsky constructs his melody. The passage begins with second violins having the higher pitch, then on the next chord, the firsts have the highest pitch. On the next, the higher note returns to the seconds, and so on. Here are the two violin parts based on their frequencies, with the upper notes in bold:

Pathétique 1

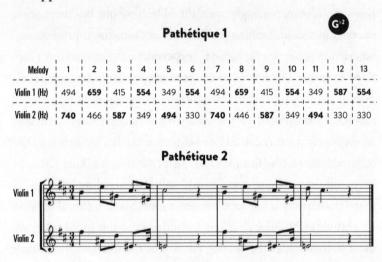

Melody	1	2	3	4	5	6	7	8	9	10	11	12	13
Violin 1 (Hz)	494	**659**	415	**554**	349	**554**	494	**659**	415	**554**	349	**587**	554
Violin 2 (Hz)	**740**	466	**587**	349	**494**	330	**740**	446	**587**	349	**494**	330	330

Pathétique 2

As the melody is completely fragmented, alternating between firsts and seconds, our aural attention is drawn from right to left to right to left and so on, as if watching a tennis match at Wimbledon. But what Tchaikovsky must have known, and what Diana Deutsch confirms, is that our brains do not perceive a fragmented melody divided across a considerable distance in a concert hall. Through previous experience and expectation of how, in general, melodies occupy higher frequencies than accompanying parts, our brains complete the jigsaw and reveal the picture. To the casual listener, the melody sounds like any other

Tchaikovsky melody. In aural terms, our Wimbledon tennis players have become one player in the center of the court. Although our right ears predominantly receive more of the first note followed immediately by our left ears receiving more of the second note, our brains interpret these as part of a connected stream of frequencies—aural connect the dots.

Why did Tchaikovsky do this? When I asked Diana Deutsch this very question, she pointed me to evidence that the conductor of the first performance, Arthur Nikisch, wanted the passage rescored so that the first violins had the melody throughout, but Tchaikovsky refused. In my experience as a conductor, the orchestral sound one hears immersed in the middle of an ensemble can be wildly different from what an audience experiences. The fragmented tennis melody surrounding Nikisch magically transformed into a perfectly functioning melodic line by the time it reached the first row of the listeners—he need not have worried; the composer was right. Will we ever know why Tchaikovsky wrote in this way? Probably not. But what we do know is that this technique still delivers a beautiful angst-ridden, passionately yearning melody, albeit through an innovative aural illusion lost on all but a very few...and you have just joined those educated few.

The Perfect Imperfection of Strings

Of all the sections of the orchestra that can deliver this level of emotional intensity, the string section is definitely the most capable (although, as a violinist and violist, I would say that). What is often overlooked, though, is that its imperfections are what make a professional string section sound so intensely emotional. Listening to any classical recordings from the 1950s onward, you will notice that the development of more sophisticated microphones and orchestral recording techniques have allowed the listener to step back from the string section, to hear it at a similar distance to that one would experience in a concert hall. Previously, microphones that had little ability to record low volume were virtually rammed down string players' throats, producing a warts-and-all sound. But by the time of Herbert von Karajan's orchestral recordings of the 1960s, the string section sounded like the smoothed-out, saccharine-filled, reverb-drenched homogenous blob we hear when we sit at the back of a large hall. Sitting *in* a professional string section, though, is a very different aural experience.

Close up, violins aren't always beautifully rounded, smooth-sounding, elegant machines capable of producing intense emotional music. They squeak, scrape, pop, grind, slide, and, even in the best hands, dare I say, often sound a little bit out of tune. There are a number of reasons for this. First, even the greatest of violin virtuosi are human, and their finger placements are accurate to only hundreds of micrometers. (If they were more accurate than this, they would sound like Cher on "Believe," which no violinist would want.) Second, all string players use vibrato, the wobbling of one's finger to give the sound more emotion. Through the physical action of rolling a finger on a string, raising and lowering its frequency (by around 6 Hz, according to researchers), every note is not a set frequency but an ever-changing pitch within microtonal tolerances. Multiplying a single orchestral violinist's slightly wobbling note by perhaps ten other first violinists means that what one hears at a distance is an approximation, an amalgam of the players' different frequencies. Add to this another ten second violins wobbling another amalgam, another eight violas with their amalgam, a further eight cellos (with even wider vibrato), and another six double basses, and one begins to understand how a professional string section has a magical, ever-shifting sheen to it, like tiny sun-drenched ripples on a calm sea. But each individual player's vibrato is not a fixed entity either. To lower emotional intensity, professional string players will instinctively slow down their vibrato, even stop it completely. And conversely, when the music requires passion, the vibrato gets bigger and grows in terms of the frequency range and speed of change; the variations are infinite. Even the greatest of orchestral string sections achieve their sound not through perfection but via imperfect frequencies.

I spend a considerable part of my working life recording string sections, but not with scores of other musicians in a posh studio. Using multitrack recording computer software, I am the cheap and increasingly popular form of string recording used for commercial music. From the likes of Coldplay's "Viva la Vida" through Clean Bandit's "Rather Be" to Dua Lipa's "Don't Start Now," record producers are using single string players recorded multiple times to create lush-sounding sections. To achieve a convincing string sound, though, one must build in imperfections and idiosyncrasies. Often, I will accidentally play a note outside acceptable tolerances (I'm avoiding using the phrase *out of tune*). But if such a mishap occurs in "take" number seven, for example, I am amazed when listening to the playback how the mistake vanishes when mixed with the others.

This form of frequency masking seems to be built into string sections through sheer safety in numbers. For example, if I record six C♯s at around 554 Hz—I suspect the parameters of acceptability are roughly between 535 Hz and 565 Hz—and my seventh is noticeably "outside acceptable tolerances," the average frequency of the majority simply drowns out the dodgy note. Indeed, if the budget won't allow for lengthy sessions with multiple retakes, I'll just play that note a little louder in the next take (and I can also, of course, call on the software of Andy Hildebrand). Similarly, I am often incredulous at how a very fast run of notes, recorded quite inaccurately over a number of takes, still sounds like an impressive grand string sweep up to a climactic moment. Just as with the Shepard effect and Deutsch's stereo melodies, our brains are often fooled by what we think we hear, or they simply invent what we're incapable of hearing. However, in any of these situations, it doesn't take many notes on the edge of these parameters to pollute the sound, at which point the client might as well have hired a school string section. Only a few moments of poor intonation can remove the veneer, and then the game's up.

What Diana Deutsch and Roger Shepard have helped us understand is that it is not our ears that are deceived but our brains. Our ears simply receive sound pressure waves at varying frequencies and act very much as the analog to digital converters in a studio, recording acoustic sound and converting it into digital information that a computer will be able to understand. It is our brains that really "hear," that are the interpretative hubs of the whole sound process. And our brains constantly try to make sense of the auditory information they receive, even if their interpretations are patently bonkers to other parts of those same brains.

Profane Opera

In 1957, American journalist Sylvia Wright published a witty article in fashion magazine *Harper's Bazaar* highlighting the human brain's desperate need to create order and sense from auditory information. It was titled "The Death of Lady Mondegreen" and tells of one of Wright's favorite poems read to her as a child,

Thomas Percy's "The Bonny Earl of Murray." Wright's memory of the opening was as follows:

> *Ye Highlands and ye Lowlands,*
> *Oh, where hae ye been?*
> *They hae slain the Earl Amurray*
> *And Lady Mondegreen.*

Wright goes on to reveal that the opening lines do not say that at all; there was no Lady Mondegreen. The actual line was "and laid him on the green," but Wright's brain didn't interpret her mother's words as that. Her brain sought some meaning, any meaning, from the rich mixture of sibilants, fricatives, vowels, and frequencies generated by human speech. Lady Mondegreen seemed as plausible as anything else—except it's not plausible in the slightest, as poor Lady Mondegreen never gets another mention in the ballad.

Ever since this article was published, misheard lines, words, and lyrics have become known as mondegreens. And to prove that this is a function of all human brains, researchers have noted that it applies not only in a vast range of languages as well as English but that it can cross between languages. I experienced a shocking musical mondegreen of my own after purchasing a CD of Léo Delibes's *Lakmé* performed by the Orchestre National de l'Opéra de Monte-Carlo, with Joan Sutherland as Lakmé. Having come to the work through the famous "Flower Duet," I happily listened through the CD, unable, though, to understand one word of the operatic French. Nevertheless, the opera was perfectly pleasant until the closing moments of act 1, scene 6, when a dramatic diminished seventh chord sets up the character Nilakantha's line, "*Il faut qu'il meure!*" which translates as "He must die!" My

brain's lack of French (I studied German throughout my school-
ing) interpreted Nilakantha's unaccompanied recitative lines *in
English* as "Ye f***ing mother!" I distinctly remember my initial
shock at hearing such profanity in an opera, as well as the sudden
change of language from French to English. I played it over and
over, trying to hear something else, but to no avail; it remained
"Ye f***ing mother!"

Perhaps more shocking than the profanity, though, is that
my brain's interpretation of the line surprised itself. It seems that
my brain's need to find sense in the line was more important than
the fact that it had created a preposterous solution. That specific
moment of the recording became my ringtone for many years. I
eventually removed it after an embarrassing incident when my
phone rang during a meeting between myself and the headmaster
of the school at which I was director of music.

That same school taught a variety of international students, and I would often entertain myself by learning phrases in
the students' first languages. Two of my party trick phrases in
Mandarin were "I'm going to my office" and "Tuck your shirt in,"
which I would often try out on my unsuspecting Chinese students. On every occasion, I would receive blank stares, followed
by my slightly more assertive repetition, followed by more blank
stares, followed by yet another repetition. Eventually, I would
give up and offer the English translation. The students, without
fail, would then nod positively and say, "Oh yes, you meant...,"
repeating back to me the exact Mandarin phrase I had just said.

What I failed to understand is that Mandarin, like Cantonese,
Vietnamese, and many other languages around the world, is a
tonal language. Simply spouting out the sounds without any reference to the intonation of the syllables and words can render
them meaningless or, perhaps more worryingly, can make them

mean something completely different. In Mandarin Chinese, there are four tones, four pitch contours for each syllable. Let's take the syllable "boar" as an example. In English, regardless of whether you mean a type of pig, an uninteresting person, a verb for digging, or the past participle of "to bear" (as in a grudge), we simply say the sound and make the listener work out its meaning from the surrounding context. However, in Mandarin, the different meanings of that same syllable are defined by the pitch and frequency contours with which it is sounded. The Mandarin four tones are: high and constant; mid vocal range and rising; low, lower, and rising (in the shape of a V); and high and falling. In another of Diana Deutsch's experiments, which explores the notion of absolute pitch, she recorded different Mandarin speakers saying a list of ten words on two separate days. When each of the pairs of words were compared across those two days, the average difference in pitch of at least a third of them was less than one quarter of a semitone. The notion that a speaker of English would always say the word *sea* at precisely the same pitch every time is very, very unlikely (unless they droned on like my chemistry teacher used to). In other experiments, Deutsch discovered that young musicians whose first language was tonal seemed to have a much higher incidence of perfect pitch.

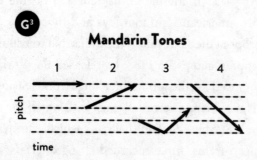

Mandarin Tones

Our Brains "Sometimes Behave So Strangely"

The jury is still out as to whether the brain's functions for speech and music are connected, but one of Deutsch's most famous experiments proves that even nontonal language speakers are attuned to frequency and pitch in their responses to speech. While working on some postproduction of her own recorded voice, Deutsch looped a small spoken phrase, "sometimes behave so strangely." Within about five or six repetitions, the phrase suddenly transformed itself into a *musical* phrase—just as it has for me and for thousands of people who have heard it since. But as with my profane line in *Lakmé*, it seems impossible for one's brain to reverse the process and rehear it in its original form. On listening back to the whole sentence, the phrase "sometimes behave so strangely" always pops out as a musical snippet, regardless of how hard one tries to force one's conscious brain to override such an aural mirage. However, as with Auto-Tune, creative people will make great art within such restrictions—and Deutsch's speech-to-song illusion is no exception.

When composer Steve Reich was only one year old, in 1937, his parents split up. Throughout the late 1930s and early 1940s, Reich visited them both, traveling by train between New York and Los Angeles. In adulthood, he reflected on the difference between his American train journeys and those of Jewish children in Europe at the same time—the trains that transported them to concentration camps. In 1988, he released the work *Different Trains*, composed for the groundbreaking Kronos String Quartet and tape ("tape" being a recorded track with which live musicians play in sync). The tape contains bells and whistles from American and European trains, three recorded string quartets, but, most importantly, snippets of speech that Reich had recorded. The speech was gathered from a variety of sources and included reminiscences from his governess and a Pullman porter who worked on the same trains during those years. In contrast, the second and third parts of *Different Trains* feature three Holocaust survivors telling of their experiences during the war and their time in the United States as émigrés post-1945.

As with Deutsch's "sometimes behave so strangely," Reich "found" the music in some of the words and phrases in his recordings, notating them in terms of both pitch and rhythm. He then surrounded these words, voices, and phrases with the persistent, mechanical locomotion of chugging violins and echoed the female words in the viola and the males in the cello. The work is an emotionally charged masterpiece, a comment on the random nature of the accident of birth and the horror of the Holocaust—he was awarded a Grammy for *Different Trains* in 1990. The genius of the work is that words and phrases are an integral part of the music and not overlaid add-ons. As the viola and cello echo and pre-echo the words, they morph the speech into song, just as in Deutsch's experiments. Whether it's Reich's governess Virginia repeatedly reciting "from Chicago to New York" on F_3 (175 Hz) and Ab_3 (208 Hz) or Holocaust survivor Rachel impassively saying "no more school" in octaves with the viola on F_3 (175 Hz), $F\#_3$ (185 Hz), and G_3 (196 Hz), each phrase initially lives in a no-man's-land between speech and music. However, our brains soon subsume them into a mesmeric and emotive all-encompassing musical sound world. It is also worth noting here a similar work that predates Reich's *Different Trains* by six years: guitarist Scott Johnson's 1980–82 composition *John Somebody* is as technically striking and novel as Reich's but at the other end of the emotional spectrum. Have a listen; it will put a smile on your face.

In my thirty years as a music teacher, I have always persuaded my students that they should "use their ears" to improve their musical skills. What this chapter's vibrational voodoo and trembling trickery has proved to me is that they should perhaps rely on their brains more than their ears. Then again, I'm not sure we should trust our brains either.

HIGH TIMES

Up into the world of ultrasound, radiation, and beyond

As with many of the subjects in this book, even the most mundane object can often hide amazing vibrational secrets. The humble microwave oven combines three applications of our understanding of frequency. To cook your frozen dinner to perfection (if that's not an oxymoron), your timer must work accurately; time itself is merely a set of vibrations. When your meal is done, a ping or beep sends an audible frequency in the upper-middle range of human hearing, telling you that it's time for your feast. But as the name of the appliance suggests, there is a third vibrational element; the food has been cooked by microwaves—in other words, radio waves.

Percy Spencer, a twice-orphaned, self-taught farm boy who became one of the United States' greatest engineers, was testing a type of vacuum tube with the deliciously sounding sci-fi name of a magnetron. He noticed that, when standing close to it, a peanut snack bar in his pocket melted. Others knew of this heating phenomenon, but Spencer's character was such that he was intrigued enough to place some popcorn in the microwave beam—and subsequently fed it to the other staff in the office. When he attempted to heat an egg, it exploded all over a colleague's face. On October 8, 1945, Raytheon—the company in which Spencer was a leading light—filed a patent for a microwave cooking oven that became known as the Radarange. Raytheon's wartime work in manufacturing magnetrons for the Allies' radar systems leads directly to your perfect two-minute frozen dinner.

In simple terms, magnetrons are like mechanized panpipes that generate electromagnetic waves. In a process that involves two magnets, a high-voltage power source, and lots of moving electrons, a wave of 2.45 GHz is fed into the heating space of the microwave oven. And that's our cue to return to chapter C and the ghost that Vic Tandy "saw" in his lab. His ghostly apparition was caused by a standing wave, a wave that reflects off a hard surface back on to itself, amplifying the points of highest energy. These antinodes (where Tandy's sword vibrated wildly) are the points where the energy of the electromagnetic waves is at its greatest, vibrating and consequently heating the water molecules in the foodstuff.

The wavelengths in a microwave oven are much smaller than those in Tandy's lab, so a collection of antinodes as well as nodes (where there is no energy at all, just as in the points on Bruce Drinkwater's levitating machine) populate the oven's space. This is why the microwave contains a turntable—it's not there to display your meal elegantly from all angles. As there are a series of hot and cold spots generated by the standing waves' antinodes and nodes, the food is moved around the oven to heat it evenly. If you've ever tried cooking without the rotating plate, you will have burned your tongue on piping hot parts of the food and then subsequently cooled it down with the completely uncooked parts. As you bite into a piece of boiling potato, even the highest scream you could possibly emit will be well short of the microwave's frequency of 2,450,000,000 Hz: perhaps singing the opening "Ah" from the aria "Ah! Belinda" in Henry Purcell's opera *Dido and Aeneas* twenty-three octaves higher would just about match it.

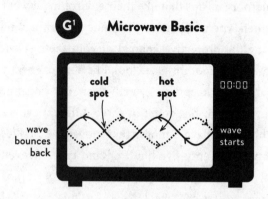

G¹   **Microwave Basics**

Such an injury moves us swiftly on to one of the most beneficial uses of high frequencies, through both sonic and electromagnetic waves: medicine. Ultrasound may not be able to soothe your burned tongue, but it does have other oral benefits.

Sound Dental Advice

One of our most direct interactions with ultrasound comes in toothbrush form. If you are the proud owner of an ultrasonic toothbrush and have been paying attention to anything in this book, you might be questioning why yours is making a noise, as ultrasound is above the upper threshold of human hearing. The answer is one of two options: you've got the hearing capabilities of a bat or, more likely, you've bought a mislabeled toothbrush.

There tends to be quite a misunderstanding when it comes to the high-tech toothbrush, so let's fill any cavities in your oral hygiene knowledge. There are manual and electric toothbrushes.

Manual ones are as good as the person using them—in the same rather primitive manner as our ancient ancestors, we are simply scraping debris off our teeth. Too little brushing, and you leave behind bacteria that becomes plaque, a form of biofilm, a community of microorganisms that like their environment wet (pond scum is another type of biofilm). Too much brushing, and you risk destroying the protective enamel of your teeth. Hence, on the face of it, electric toothbrushes look like a more effective and controllable means of keeping our teeth clean. But rather confusingly, electric toothbrushes come in three different types. Basic electric toothbrushes have a motor that moves the bristles back and forth between twenty-five hundred and twenty-seven hundred strokes per minute. Though it would be easy to mock our increasing twenty-first-century laziness at this point, electric toothbrushes have very important health benefits for those with motor control issues, for example. There is also evidence that the thousands of strokes per minute they produce are more efficient than our mere manual scrubbing of about three hundred.

Next in the trio of plaque-based perplexity is the sonic toothbrush. Sonic toothbrushes are, as their name suggests, ones we can hear (even though we can also hear the electric ones). Their stroke rate is considerably higher, though, at around thirty thousand strokes per minute, which works out at about middle C on the piano at 261 Hz. Because the bristles of the brush move so much faster than those of a conventional electric toothbrush, it would seem reasonable that it would clean more quickly and efficiently. But in addition, manufacturers suggest, fluid dynamics are at work on the biofilms in our mouths, due to the high-intensity vibrations of the middle C. This secondary process uses water and saliva to help dislodge bacteria, through both pressure waves and bubble formation. The sound it produces comes from the motion of the motor that turns the bristles and is not, for the most part, responsible for the cleaning of the teeth (though many claim that the C does have some impact).

There are two ways in which you can tell if you're using the third in our toothbrush trio. First, it is silent, as it works in the frequency range above human hearing. Second, you will have paid a considerable amount of money for it. The ultrasonic toothbrush works using a principle called cavitation. It is a process by which voids or bubbles in a liquid collapse under pressure. This in turn produces a shock wave, which can potentially be damaging. Cavitation is a destructive force in many areas of engineering, from the blades of tidal wave turbines and ships' propellors to the internals of diesel engines. It has also been suggested that the 1883 eruption of Krakatoa was the sound of cavitation—where giant bubbles in the magma combined and collapsed, with sonic consequences for the whole planet. But cavitation is also used to great benefit in nature. Ferns use cavitation to release and catapult their spores across a wide area. And the same process is exploited by the pistol shrimp to kill its prey (though this could be seen as a negative application of cavitation if you happen to be a small fish in the vicinity). Returning to our toothbrush, its frequency of around 1.6 MHz acts on microscopic bubbles in water and saliva, causing them to collapse, sending shock waves through liquid to the surface of your teeth and gums. This frequency, 1,600,000 Hz, is thirteen octaves up from the opening phrase of Barry Manilow's "Can't Smile without You," which weaves around the note G_3 at 196 Hz—and though there is no evidence that Manilow used an ultrasonic toothbrush, he once revealed during a radio interview that he brushed his teeth every two hours. In theory, the shock waves from these violent cavitation collapses dislodge any pond scum that has attached itself to the surface of our teeth. While research indicates that there is a significant increase in plaque removal using electric toothbrushes over manual ones, the evidence suggesting that cavitation effects alone can clean teeth is a little scant. As with most things in life, it seems that a combination of factors and processes (ultrasound shock waves *and* brushing, in this case) is the most effective solution.

Seeing through Ultrasound

Perhaps the most widely known medical application of vibration is that of the ultrasound scan. Following the successful use of high-frequency radar during the Second World War, many progressives in the medical profession wondered whether the technology, used to find enemy aircraft in the skies, could be turned in on ourselves to seek out lumps and bumps in the human body. During the 1950s, rudimentary ultrasonic scanning devices began to emerge as useful diagnostic tools, and the discipline of sonography was born. Though sonographers can now scan large parts of our bodies, the areas most commonly associated with ultrasound scans are the heart and the womb—especially when there's a baby in it.

The birth of obstetric ultrasound resulted from a light bulb moment: not a metaphorical light bulb, an actual light bulb. In 1956, engineer Tom Brown was fitting one at Glasgow's Western Infirmary Hospital when he overheard obstetrician Ian Donald discussing problems with the hospital's diagnostic machines. Within two years, the medical skills of Donald and the engineering talents of Brown—along with obstetrician John MacVicar—were combined in a paper published in *The Lancet* titled "Investigation of Abdominal Masses by Pulsed Ultrasound"; it included the first ever ultrasound images of a fetus. Though their prototype ultrasound scanner was a mishmash of spare parts (including pieces of Meccano and an industrial scanner used for finding flaws in metals), within a short time, Glasgow was leading the world in obstetric and gynecological ultrasound scanning.

The principle of ultrasound scanning is the same as radar or bats' echolocation. A very high-frequency sound is transmitted at an object, and the reflected waves that bounce off it are received back at the source and translated into something meaningful, whether that be supper for a bat, an approaching enemy bomber plane, or the image of a fetus. In their 1958 paper, Donald et al. explained how wave reflections are generated from the interfaces between two materials of differing densities; the larger the difference, the more reflection occurs. But they also mention such pitfalls as cavitation—nobody wants potentially destructive microscopic bubbles imploding inside them—and the production of heat when too much sound is absorbed.

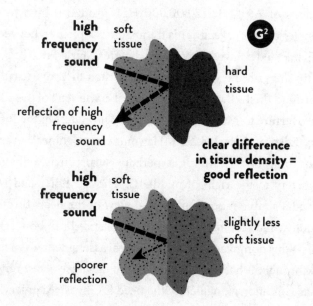

Earlier in the decade, another team of visionary minds was working on ultrasonic scanning of the heart, and this team had quite a pedigree. German physicist Carl Hellmuth Hertz (the great-nephew of the great Heinrich Hertz), alongside Swedish cardiologist Inge Edler, developed echocardiography in 1953. Edler was particularly interested in mitral valve problems. The heart's mitral valve allows reoxygenated blood to flow from the left atrium down into the left ventricle before being pumped out on its journey around the body. He was keen to assess problems *before* cutting open his patients (and so were his patients, one suspects). Just like Donald and Brown, Edler and Hertz's first machine was a rather Rube Goldberg affair; they borrowed an industrial ultrasonic reflecto-scope from a local shipyard near Lund in Sweden. The machine had been developed for detecting flaws and cracks in shipbuilding materials, but when used on patients, Edler found he could identify potential blood flow problems associated with the mitral valve.

They landed on an optimal ultrasonic scanning frequency for adults of 2.5 MHz (2,500,000 Hz), produced by a twelve-millimeter quartz crystal. This frequency is a pitch in between D and E, thirteen octaves above the two notes Marie Fredriksson of Roxette sings on the word "heart" in "Listen to Your Heart"; it's also around the E♭ that the violins repeat at the start of the allegro of the overture to Mozart's opera *The Magic Flute*. On the Infinite Piano, Edler and Hertz's 2.5 MHz (modern machines can transmit up to around 18 MHz) is, perhaps surprisingly, only about two meters to the right of middle C. Though Edler and Hertz received a simple monodimensional view of the heart through a readout—which looked like a slice of collapsed rainbow cake in profile—this noninvasive form of medical assessment was a major breakthrough. It has led to our ability to diagnose a vast range of cardiac conditions long before the need for invasive surgery.

But there is a bizarre twist in this story, a metaphorical swerve in the road of technological progress that few could have predicted. Carl Hellmuth Hertz was not content with collaborating on an invention that would one day save the life of my unborn daughter among countless others. As Hertz needed a better way of recording the readouts from those early machines, he was one of the leading developers of the inkjet printer. Let's pause for a moment to inwardly digest that. As well as playing a pivotal role in the development of modern echocardiography, Carl Hertz was at the forefront of the technology now commonplace in all our homes and offices—the printer.

Though, like the microwave, this may seem a rather humble invention when compared to a machine that can "see" inside our bodies, the scientific knowhow in a modern inkjet or bubble-jet printer is quite astounding. In his continuous "ink jet recorder" patent of 1968, Hertz and collaborator Sven-Inge Simonsson explain a number of different means of continuous inkjet printing, all of them based on the principles of bubble formation and liquid surface tension postulated by nineteenth-century scientist Lord Rayleigh and Marcel Minnaert, whom we met in chapter C. Hertz and Simonsson noted that a jet of liquid (in this case ink) could be changed into individual droplets by adding an electrical charge to the flow: "the recording trace on the paper can be interrupted by applying the control voltage...and it can be shown, that this can be done up to 1,000,000 times per second [a frequency of 1 MHz]." But the truly ingenious part was that the individual drops from this continuous flow could be fired at the paper *or* diverted and recycled. All the droplets pass through a small electric field that is placed between the nozzle and the paper. The unwanted droplets, which are electrically charged by the control voltage, are deflected by the electric field and

collected for reuse. This means that only the uncharged droplets of ink land on the paper. Hertz and Simonsson showed they
could precisely control the ink at any given point. Using an array
of these nozzles to enhance speed, color, and accuracy, Hertz's
continuous inkjet printing was his second major scientific con-
tribution to society.

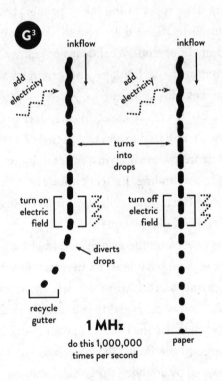

At present, there are several different ways of generating
droplets, but the speed and miniaturization of this technology is
truly staggering—printheads can have over forty thousand nozzles on them, each around fifteen microns tall, which is smaller
than the width of a human hair. And this level of accuracy is
now leading inkjet printing far beyond the reproduction of our
favorite family photos, indeed full circle back into the world of

medicine. If such machines can deliver minute droplets of liquid with incredibly consistent levels of accuracy, why limit the process only to ink? Inkjet printing has incredible future potential for the delivery of individualized medication. And a recent article in *Nature* highlights the extraordinary work of researchers who have printed solar panels on to the surface of soap bubbles—the potential ability to power lightweight, flexible devices such as medical patches is only just around the corner.

Ascending to Bikini Snow

Though high-frequency mechanical waves are mostly seen as beneficial and benign (although this isn't always the case), electromagnetic waves—usually known to the layperson as radiation—strike fear into us all...and at certain frequencies, with good reason.

In the lower part of the electromagnetic spectrum, waves are known as nonionizing radiation. In other words, they do not have enough energy to change the state of atoms and molecules. As the frequency of electromagnetic waves increases, they energize atoms and vibrate molecules, resulting in an increase in temperature. But all nonionizing radiation lacks sufficient energy to destabilize the equal number of electrons and protons in an atom.

Playing an Infinite Piano scale of radio waves from the lowest note in the universe up past middle C (and for at least another twenty octaves beyond) would result in a gradual increase in the heating of particles but no change in the state of atoms and molecules. Around twenty-four octaves above middle C, one can find microwave frequencies whose energy is enough to heat water molecules, allowing that frozen dinner to cook in the oven. Continuing up the electromagnetic scale to about thirty octaves, one would now be playing the lower frequencies of infrared radiation—between 300 GHz (300,000,000,000 Hz) and 400 THz (400,000,000,000,000 Hz). These are the "notes" of the sun and fire, the "sound" of which warms us.

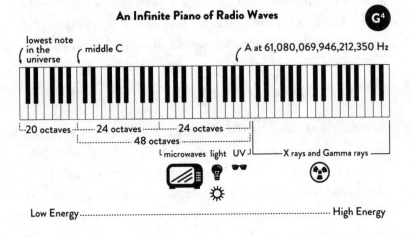

An Infinite Piano of Radio Waves

Add another eleven octaves (we are now around 6.7 meters to the right of middle C), and we encounter the single octave of frequencies that are visible to us, light. Even at this point, radiation is still nonionizing. Although putting one's head in a microwave oven or being burned with fire is obviously very damaging, such damage is only tissue-based—it does not lead to the corruption of cell regeneration and cell death associated with ionizing radiation.

Continuing along the Infinite Piano, up into the next octave, we leave behind the visible spectrum and begin to play ultraviolet light. The boundary between nonionizing and ionizing electromagnetic radiation now begins to blur, and as we near the forty-eighth octave above middle C, a health warning suggests we should approach the Infinite Piano's keys with great caution (just around an A at 61,080,069,946,212,350 Hz). From here on up, the energy of electromagnetic waves is strong enough to split electrons from their orbits around host atoms, leaving unequal numbers of electrons and protons in a single atom—which is now called an ion.

Electrons can be found in pairs in the orbits of atoms. Atoms, or ions, with an unpaired electron are called free radicals. The lonesome "odd" electron found in free radicals goes in a desperate search of other things to bond with, and this is where the trouble starts. Messing with our atoms by changing their states into ions and free radicals is not a good idea. Human exposure to such ionizing radiation plays havoc with our bodies at the smallest level, damaging molecules of DNA, proteins, and carbohydrates. As we climb higher and higher up the Infinite Piano, we sound increasingly dangerous notes through the X-ray and gamma ray octaves.

On March 1, 1954, twenty-three Japanese fishermen got far too close to these lethal notes. They were fishing for tuna in the waters around the Marshall Islands, a group of atolls situated in almost half a million square kilometers of the Pacific Ocean. Led by an inexperienced captain, twenty-two-year-old Hisakichi Tsutsui, the crew of the *Daigo Fukuryū Maru* had had a tough time since setting out from Japan in late January. After losing almost half their fishing lines in February, they ventured much farther south in search of bigeye tuna and found themselves approximately 130 kilometers east of Bikini Atoll. Since 1946, the U.S. government had been using the atolls of Enewetak and Bikini (among others) to test and advance its nuclear weapons program. The Americans had established a military danger zone around Enewetak, which Tsutsui knew about, but he was unaware that the zone had been expanded to include Bikini. Nevertheless, the *Daigo Fukuryū Maru* was still just outside the zone on March 1—albeit by quite a short distance—and was not noticed by any spotter planes or picked up on radar.

Castle Bravo was the code name for the thermonuclear weapon that was detonated in the northwest corner of Bikini Atoll on that morning. The bomb was one thousand times more powerful than the one dropped on Nagasaki only nine years previously, and it remains the largest nuclear weapon ever detonated by the United States. Its power was equivalent to fifteen million tons of TNT, and the explosion was much larger than the American scientists and military had anticipated. Though the blast was startling ("the day the sun rose in the west," as fisherman Ōishi Matashichi described it), neither boat nor crew were damaged. Later, though, "Bikini snow" started to fall: rain that contained a white ash soon covered the boat, the nets, the catch, and the crew. They spent a number of hours hauling in their lines in order to get away, attempting to clean up and bag the ash with their bare hands—Matashichi even took a lick of the ash to try and establish what it was. Some of the crew also collected it in small vials and pouches as souvenirs, which they hung on their bedposts down below.

Unbeknownst to them, Bikini snow was highly radioactive fallout (which became known in Japan as *shi no hai*—"ashes of death"), a lethal mix of contaminated coral, sand, and dust that included radioactive isotopes such as cesium-137, selenium-141, strontium-90, and uranium-237. Using these isotopes' lowest level energy state wavelengths as a basis for frequency calculations, it is possible to roughly estimate that the death ash notes, transposed lower, would sound a rough chord of F major in first inversion: uranium as bass on a slightly flat A_3, strontium on the tenor part as a slightly flat A an octave above, selenium as the alto on a slightly flat C_4, and cesium acting as the soprano note around F_5. The sound of these notes is not dissimilar to the dramatic final chord of Giacomo Puccini's opera *Madama Butterfly*, a work in which, rather prophetically, an American betrays his innocent Japanese wife in a reckless act that results in her death, set on a hill overlooking Nagasaki. The Castle Bravo detonation happened almost exactly fifty years after the premiere of Puccini's masterpiece.

The lethal Bikini snow penetrated the eyes, noses, and mouths of the crew—they were contaminated by the ash, as they were in direct contact with it (this is different from being irradiated, where one is exposed to the electromagnetic waves emitted by radioactive material). Within hours of the fallout landing on their boat, the crew were vomiting and suffering from dizziness and fevers. By the time they arrived back at the port of Yaizu two weeks later, they were in a very sorry state, having suffered skin burns, bleeding gums, and hair loss. The horror of the atomic bombings of Hiroshima and Nagasaki was still very raw in the Japanese psyche, and it was understandable that the national press covered the story of the fishermen so closely. News of the incident soon spread throughout Japan, including reports that the crew's catch of radioactive fish had been sold throughout the country. The ensuing panic brought the Japanese fishing industry to its knees.

Through study of the crew of the *Daigo Fukuryū Maru*, experts soon realized that the survivors of the Hiroshima and Nagasaki explosions (*hibakusha*) had become ill not only from radiation emitted in the initial blast but from the radioactive material that rained down on them in the following hours; they called it "fallout." The unlucky crew of *Lucky Dragon 5* (as *Daigo Fukuryū Maru* translates into English) were suffering from the newly named "acute radiation sickness." The electromagnetic waves emitted by the radioactive isotopes in Bikini snow were ionizing radiation, with enough energy to separate individual electrons from their respective atoms. And if those atoms happen to be part of a human body, then all hell can break loose—our very DNA begins to generate abnormal cell divisions, leading to cancers and a host of other medical problems.

For the crew of the *Lucky Dragon 5*, the nightmare was just beginning. Many of the twenty-three developed cancers over the years, and a number died of cirrhosis of the liver. Ōishi Matashichi suffered from liver cancer as well as the trauma of losing a child who was stillborn. The social impact of the *Lucky Dragon 5* incident was considerable. The Japanese (and eventually people around the world) found their oppositional voice to nuclear weapons—it was this event that sparked the beginning of the antinuclear movement, not the bombings of Hiroshima and Nagasaki. And the character of Godzilla was unleashed on the cinema-going public within a year of the detonation of the Castle Bravo bomb. This terrifying creature, awakened from the depths of the Pacific Ocean by nuclear radiation, was a direct and challenging metaphor, highlighting the threat posed by such weapons.

Following the Castle Bravo detonation, the inhabitants of the Marshall Islands were also exposed to radioactive fallout, even though they had been evacuated from their homes to "safer" islands. As Lewis Strauss, chairman of the U.S. Atomic Energy Commission, put it in a press conference in 1954, "the wind failed to follow the predictions." Was he suggesting any failure to protect the inhabitants was the wind's fault? This breathtaking evasion was part of a wider U.S. government strategy, perhaps summed up in a simple phrase: "What bomb?" The Americans had tried to keep their atomic weapon tests secret, though one would have thought that exploding thermonuclear devices is a little difficult to hide. The *Lucky Dragon 5* incident marked the end of this secrecy. The unfortunate Marshallese people from Enewetak and Bikini, displaced prior to the detonations, were now contaminated by fallout because the wind had failed to blow in the correct direction. And to add insult to considerable injury, it took days to evacuate them from these "safer" islands following the explosion.

In 1958, the U.S. government ceased nuclear testing in the Marshall Islands, and for many, this drew to a close the most terrifying chapter to date in humanity's sad eternal book of war. However, the Marshall Islands and their people continue to live with the very present threat of these radioactive frequencies. In a recent investigative article, researchers reported that "we find highly elevated gamma radiation levels on Bikini Island." Of particular concern is the island of Runit in the Enewetak Atoll. Between 1977 and 1980, the Americans grudgingly embarked on a "cleanup" of the radioactive debris from the tests around the atoll. Using one of the original blast craters on Runit Island, four thousand unprotected U.S. servicemen dumped an estimated seventy-three thousand cubic meters of contaminated debris and topsoil from the atoll's islands into the crater, covering it with a forty-five-centimeter-thick concrete dome. The report continues by saying that "the presence of radioactive isotopes on the Runit island is a real concern, and residents should be warned against any use of the island." The dome now shows signs of breaking up, and rising sea levels means that these isotopes are being washed back into the lagoon and wider ocean. The Marshallese who moved from their homes back in 1946 were told they were doing so "for the good of humanity"; many have yet to return. If they were to go home, the potential harm from background levels of high-energy electromagnetic radiation, along with the danger of contamination through ingesting foods grown in the soils, could prove to be a lethal cocktail of risks. They continue their struggle for a comprehensive cleanup and fair level of compensation for the damage wrought on these remote and beautiful islands.

The Highs and Lows of X-rays

It is true that the high frequencies of electromagnetic waves we call ionizing radiation are very dangerous. However, fire is dangerous, but humans still utilize the radiation it emits to positive effect. For whatever reasons, our society views these superhigh electromagnetic frequencies almost exclusively with great suspicion and fear. We counter our view of the danger of fire through our ability to understand and control it—we can extinguish it when we want (mostly), we know how close to get to it before it damages us, and we have learned to harness it to keep warm, cook our food, and make things. However, our society tends not to have such a balanced view of ionizing radiation. For most of us, radiation means Chernobyl, Hiroshima, cancer, and certain death, a one-dimensional reaction to a phenomenon we encounter every second of every day.

There are three types of radiation (alpha, beta, and gamma), all of which have differing levels of danger. Alpha particles are the most damaging to human cells but can be easily blocked by skin, paper, or other thin materials; they have poor penetrability. However, the ingesting of isotopes that release alpha particles into the body is highly dangerous. Beta radiation is also very harmful, but its penetration potential can be countered by materials as thin as an aluminum sheet. Gamma rays are the least damaging but the most dangerous overall because their penetration is extremely difficult to counter.

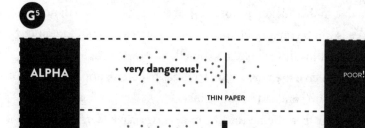

RADIATION	MATERIAL	PENETRABILITY

Even when armed with this knowledge, it can still prove difficult for us to make rational judgments about what is dangerous and what is not. In 2019, BBC science correspondent Victoria Gill reported from Chernobyl, "I can see the nuclear power plant—less than one kilometer away from the reservoir bed we're standing on." She was accompanied by Professor Jim Smith of University of Portsmouth with his useful radiation-measuring gadget, a dosimeter. "It's currently reading 0.6 [microsieverts per hour], so that's about [a third] of what we were getting on the flight." Though there continue to be many hot spots and no-go areas around Chernobyl, it will probably come as a surprise to learn that we would receive more gamma ray radiation on the airline flight to Chernobyl than standing for the same amount of time only one kilometer from the nuclear plant. On the plane to Russia, Gill and Smith would have been subjected to electromagnetic waves from the cosmic rays and the sun that we, eleven thousand meters below on the ground, receive some protection from due to the thicker atmosphere at sea level. It is worth noting that cells can be repaired if the amount of damage is not too much or too prolonged; for lighter-skinned people, a week's holiday in the sun can give them a suntan, whereas a lifetime unprotected in that same sun can lead to skin cancer.

But far from being uncontrollable, we have learned to harness the considerable potential of ionizing radiation. The lethal power of gamma rays is of great therapeutic use in our treatment of cancer, and our ability to control X-ray radiation has given us the diagnostic capability to see through the human body. On November 8, 1895, physicist Wilhelm Röntgen was experimenting in his lab in Würzburg, Bavaria, with a cathode ray tube, a kind of giant light bulb. It had a negatively charged cathode at one end and a positively charged anode at the other, both hooked up to a high-voltage electrical source. When the apparatus was fired up, it glowed a bluish-green, resulting from electrons being attracted at great speed from the cathode to the anode. Röntgen was convinced—just as his contemporaries Heinrich Hertz, Nikola Tesla, and Philipp von Lenard also were—that unseen rays were being emitted from the cathode ray tube. Röntgen put a box over the tube, stopping any of the bluish-green light from escaping. After he turned on the apparatus, he happened to notice out of the corner of his eye that a screen of card he had covered with barium platinocyanide (in preparation for his next experiment exploring fluorescence) flickered and glowed—it was a couple of meters away from the tube. He turned the tube off, and the screen went dark. He turned it back on again, and the screen's glow returned. Röntgen was now sure that a mysterious wave was traveling from the tube to the screen. As it was unknown, he gave it the mathematical symbol X.

Over the next month and a half, Röntgen lived, ate, and

slept in his lab, experimenting by placing objects in between the tube and the screen. Some things left a dark shadow and were obviously impenetrable by his "X-rays"; other things, such as paper, allowed the ray through and onto the screen. Röntgen then replaced the screen with a photographic plate and "volunteered" his wife's hand as the subject for the world's first radiograph. "I have seen my death," she is claimed to have said when she saw the bones in her hand. Nevertheless, after the publication of his work "On a New Kind of Rays" on December 28, 1895, the world went wild for X-rays. For decades, fluoroscopes were an entertainment attraction, and children's shoe-fitting fluoroscopes were still in operation in some department stores in the 1970s. There is little Bob trying on some new shoes, placing his feet in the machine; the shop assistant, parent, and child peer through the viewfinders...and all three are subjected to a blast of X-rays. "I don't like these shoes," little Bob says, so he tries on a different pair and all three get another dose of radiation. "Mmm, I can see that the bones in little Bob's right foot seem a little close to the top of the shoe," the shop assistant helpfully points out. "Let's try the next size up... Come on, Bob. Let's have another look through the fluoroscope to see if they fit better." The warm glow that little Bob, Mother, and shop assistant feel after their success-ful shoe purchase might not have been solely an emotional one. It is undoubted, though, that Röntgen's discovery has changed the world, allowing clinicians to peer within our bodies, airport secu-rity to peer in our suitcases, and the Chandra X-ray Observatory to peer at that lowest note in the universe.

Saving Time

Frequency reaches into every part of our lives, our planet's life, and our universe; it even dominates time itself. Indeed, time is merely yet another set of vibrations—and there are a number of different frequencies we use to measure it.

Almost single-handedly, business executive Nicolas Hayek saved the Land of Time. As the last seconds ticked by toward the almost inconceivable end of an era, Hayek managed to rescue the Swiss watchmaking industry from oblivion. Watch and clock making had always been to Switzerland what haggis is to Scotland. Until 1969, it seemed that the exquisite quality of Swiss-made watches would guarantee the country's world domination of wrist-borne timepieces forever. However, a precocious young upstart of a company called Seiko had the temerity to challenge the Alpine purveyors of precision. Well, they weren't upstarts exactly: Seiko was actually established in 1881 by a twenty-one-year-old Japanese entrepreneur named Kintaro Hattori. On Christmas Day 1969, Seiko launched the Astron watch, triggering the quartz revolution.

Though it was the Japanese who named it the quartz revolution, the Swiss understandably dubbed it the quartz *crisis*. Until Christmas Eve 1969, it was assumed—by the Swiss—that the time on our wrist would always be driven mechanically; a spring wound full of energy, released at regular intervals by a series of almost microscopically engineered gears and cogs, creates seconds, minutes, and hours. This precision meant that, without any help from batteries or electricity, a mechanical watch loses only a few seconds of time per day. The Seiko Quartz-Astron didn't try to compete with the Swiss tradition of quality mechanical engineering. Instead, they sold space age accuracy using the technology of vibration—the new quartz watches were accurate to around 0.2 seconds per day, or one minute per year. The Swiss were caught off guard. They had released a quartz prototype Beta 1 in 1967, but their manufacturing systems were not geared up for a rapid transition to this "new" crystal. A ball made of such a crystal might have helped them see that this crisis was a long time in the making.

Back in 1880, French brothers Jacques and Pierre Curie discovered that quartz had piezoelectric properties. Some crystals—including quartz—generate tiny amounts of electric energy when mechanical stress is applied to them. And conversely, they change shape and oscillate when electricity is passed through them. The first quartz clock was built in 1927, and by the 1930s, these rather cumbersome and incredibly expensive timepieces were already accurate enough to be measuring variations in the earth's rotational spin.

If cut, sized, and shaped correctly, quartz vibrates at 32,768 Hz when stimulated by an electrical charge (this is a slightly flat C, seven octaves above middle C at 256 Hz, that frequency with "magic" powers we encountered in chapter B♭). At first glance, this ultrasonic frequency seems to be a bit of a random number. However, 32,768 is a power of two, meaning that if one halves it and halves it again, etc., after fifteen divisions, one arrives at the number 1. In the case of vibrating quartz, repeatedly halving its vibrations to 1 Hz gives a measurement of one vibration per second. Picture an old-fashioned water mill with 32,768 paddles completing one revolution every second, one paddle of which is made of metal and not wood. If a bell is placed on the wheel, there would be just one metallic ping of the bell every second as the metal paddle struck the metal bell. Although the principle behind quartz timing is not wheel- or bell-like, hopefully you get the idea.

There are two other notable facts about quartz and its use in time: first, that the shape of the crystal used inside a watch is that of a tuning fork, something that would undoubtedly make John Shore, the tuning fork's inventor, very proud (though it's so small, he'd struggle to see it); and second, quartz is one of if not *the* most abundant mineral on earth—great news for potential mass production. By the early 1980s, the Swiss mechanical watchmaking industry was on its knees, until Nicolas Hayek called time on the old practices and disjointed manufacturing and marketing processes. After a series of mergers, a lot of quartz, a whole heap of plastic, and a clever advertising campaign aimed at hip young time tellers, the brand Swatch saved a vital piece of the Swiss national identity.

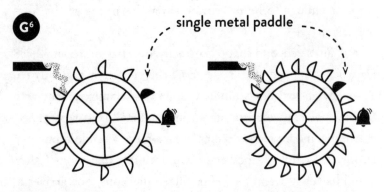

more paddles = more accurate placement of the metal one

Even before the public clamor for quartz accuracy (of a loss of no more than one second per week), scientists were aiming their sights much higher—to within one second every one hundred million years. To achieve such precision, a tiny tuning fork–shaped piece of quartz was never going to vibrate fast enough to enable the calculation of a true second. A water mill wheel with 32,768 paddles leaves too many gaps in between—scientists needed a wheel with billions of paddles, enabling that single metal paddle to ping the bell with pinpoint accuracy.

Keeping Time

To find materials with such fast vibrations, researchers and inventors turned to much smaller states of matter—atoms. Even as far back as 1879, Sir William Thomson (Lord Kelvin) and Peter Guthrie Tait predicted in their book *Treatise on Natural Philosophy* that "it is possible that at some not very distant day the mass of such a sodium particle may be employed as a natural standard for the remaining fundamental unit." Kelvin and Tait were searching for a unit of time that did not rely on the earth's rotation, as, believe it or not, this is variable *and* decreasing (some suggest the spin of the earth has slowed by about six hours in the last 2740 years). A few paragraphs earlier in their treatise, they talk of the vibration of quartz as a possible "absolute unit," in that it "gives us a unit of time which is constant through all space and all time, and independent of the earth."

A super accurate clock that can measure time throughout the universe while remaining unrelated to the earth's spin was unveiled in 1955. Built by Louis Essen and Jack Parry at the UK's National Physical Laboratory in Teddington, Middlesex, the clock led to a redefining of the unit of time of a second to such accuracy that atomic time and the rather slack earth time began to diverge. To counter this, we now add leap seconds every so often to realign ourselves with the inaccuracies of our planet's rotation. When this happens, the coordinated universal time (UTC) of carefully maintained atomic clocks is stopped, added to, or smeared so that it can be synchronized once again with universal time (previously known as Greenwich mean time), which

NASA says is "based on the imaginary 'mean sun' which averages out the effects on the length of the solar day caused by the Earth's slightly non-circular orbit around the sun." But these are not the only types of time: there is also international atomic time (TAI), which is a form of average of over two hundred atomic clocks around the world; global positioning system (GPS) time, used by satellites, which is offset from TAI by around nineteen seconds; and terrestrial time, which is measured by astronomical observations, is over thirty seconds offset from TAI, and is often deemed to be "theoretical." When added to system time, sidereal time, and standard time (which is adjusted with daylight saving time), one can see why time is a rather complicated affair.

But Louis Essen's atomic clock, the realization of Lord Kelvin's prediction, shows just how far science progressed in only eight decades. How does it work? Here's the basic idea. The resonant frequency of each different type of elemental atom is constant, regardless of where one is in the universe. Such a constant is most helpful when attempting to measure anything. The resonant frequency of an atom is detected when it is bombarded with radio waves at the same frequency. It absorbs some of the energy of the radio waves and changes to a higher energy state (in the same way that the Tae Bo Twenty-Three, who happened to be exercising at exactly the Techno Mart tower's resonant frequency, increased the skyscraper's energy state). A bunch of cesium-133 atoms, their resonant radio wave frequency, and the humble quartz crystal, all bundled in a loop, allow an atomic clock to remain self-adjusting, keeping the atoms oscillating at precisely 9,192,631,770 Hz. This frequency was adopted as the basis for the SI unit of a second in 1967.

It is extraordinary that even time itself can be distilled down to a vibration, albeit a very fast one. This frequency, 9,192,631,770

Hz, sounds a very high C♯, twenty-five octaves above the final note that Debbie Harry sings at the end of Blondie's song "Atomic" (on the Infinite Piano, you would have to travel four meters to the right of Harry's C♯ key). The ability to split one second into over nine billion subdivisions is pretty useful for measuring time so accurately that such a clock only loses one second every three hundred million years. But cesium's accuracy is a mere guesstimate of a measurement compared to the precision of strontium atoms, which emit radiation at 429,500 billion Hz—thirty-nine octaves (or 6.4 meters on the Infinite Piano) above the G_5 the celesta plays at the start of the "Dance of the Sugar Plum Fairy" from Tchaikovsky's ballet *The Nutcracker*.

But who really cares whether their alarm goes off at 6:30 a.m. or perhaps 6:30:002 a.m. the following week? Is it *that* important? Such questions have always been asked of time, whenever humans have been required to conform to ever-greater standardization. As railway travel became an increasingly essential mode of transport in the UK during the 1840s, opposition to synchronizing time was rife. Local time throughout the towns of the UK—often based on a sundial—was not the same as that in London (for example, Bristol was eleven minutes slower). It is hard for us now to imagine the indignation felt by locals that time decreed by the sun was somehow "wrong" and that a travel company based in London called the Great Western Railway should "correct" this, dictating when the local population should get up, go to bed, or get on a train.

Clocks in Oxford and Bristol, among others, had two minute hands, one for the local time and one for the newly imposed Greenwich mean time. But time standardization was about more than local pride. In August 1853 at Valley Falls in Rhode Island, two trains collided in a horrific head-on crash. As the line between Providence and Worcester was single track, the trains had to rely on timing to pass safely. The southbound train was two minutes late, and the resulting picture of the wreck (the first ever photograph of such a crash) undoubtedly spurred the authorities to standardize time across the network. The *New York Times* reported the incident rather euphemistically as follows: "The collision occurred between the regular up train and the excursion train from Weeting's. The latter was out of time, and met the up train at Valley Falls." Standardized time wasn't only a matter of knowing when your train left the station; it could be a matter of life and death after you had caught it.

The End of Time

Nowadays, standardized GPS time is essential for flying our planes, trading across international markets, and even directing us to our child's new gym class location via our cell phone. In the future, time accuracy will also be vital for our more audacious travel plans. If astronauts were to rely on quartz crystal clocks for a trip to Mars, they could lose about a millisecond for every six weeks of spaceflight. Though this seems a mere fraction of time hardly worth quibbling over, landing at a predestined location on Mars after a seven-month journey would place them about fourteen hundred kilometers off target—a very long walk back to the Martian base in a cumbersome spacesuit.

I shake myself from my Martian daydream and peel one of the radioactive bananas in my fruit bowl. I need a snack to keep my energy up—writing is hard work! The out-of-tune start-up chime of my Mac computer disturbs my purring cat, who is sunning herself in the light rays that stream in through my window. I glance out to see a rainbow beyond the end of my garden, my small patch of land full of buzzing bees, undoubtedly the odd rat, and innumerable glistening spiderwebs resonating from the struggles of trapped insects. Overhead, a flight of pigeons circle, the sound of their collective wingbeats drowned out by my daughter's piano practice coming from the next room...sounds like Tchaikovsky's *Pathétique* Symphony; I must get that piano tuned. I hear the ping of the microwave oven downstairs, my cue to head down for supper. The smell of kung pao chicken fills my nose, and I salivate at the thought of a spicy meal full of Szechuan pepper. In the kitchen, "Good Vibrations" is playing on the radio. I glance at my classic Swiss-made quartz watch—eight o'clock, give or take a few seconds. I'm exhausted after a busy day at work, and I can feel my brain slowing as the sun sets. After the meal, I head back upstairs to prepare for bed. I brush my teeth with my new ultrasonic toothbrush as a police siren wails outside, and I'm engulfed with a sense of doom-laden fear. What's that out of the corner of my eye? I must get that faulty extractor fan fixed. Getting into bed, I welcome the peace and quiet, the escape from noise. Sometimes, it can be overwhelming, you know, what the ear hears...and doesn't.

SOURCES

Here is a quick guide to some of the sources I have used as well as suggested further reading.

Ab—Prelude

two-tone mix-up: R. V. Jones, *Most Secret War* (London: Hamish Hamilton, 1978), 98.

exasperated by the failure: Jones, *Most Secret War*, 151.

red light is a form of electromagnetic wave: Dr. Barbara Mattson, "Imagine the Universe!," NASA Goddard Space Flight Center, updated November 5, 2013, https://imagine.gsfc.nasa.gov /science/toolbox/spectrum_chart.html.

the purr is a self-healing device: Elizabeth von Muggenthaler, "The Felid Purr: A Healing Mechanism?," *Journal of the Acoustical Society of America* 110, no. 5 (November 2001): 2666, https:// doi.org/10.1121/1.4777098.

a recent study concluded: Shengwei He et al., "Low-Frequency Vibration Treatment of Bone Marrow Stromal Cells Induces

Bone Repair *in Vivo*," *Iranian Journal of Basic Medical Sciences* 20, no. 1 (January 2017): 23–28, https://doi.org/10.22038 /ijbms.2017.8088.

bananas are radioactive: Joe Schwarcz, "Is It True That Bananas Are Radioactive?," Office for Science and Society, McGill University, March 15, 2018, https://www.mcgill.ca/oss/article /you-asked/it-true-banana-radioactive.

A—The Black Hole at the End of my Piano

the note was a B-flat: Brian Dunbar, "Interpreting the 'Song' Of a Distant Black Hole," NASA Goddard Space Flight Center, November 17, 2003, https://www.nasa.gov/centers/goddard/universe/black _hole_sound.html.

Earth's minimum and maximum orbital velocities: David R. Williams, "Mercury Fact Sheet," NASA Goddard Space Flight Center, updated December 23, 2021, https://nssdc.gsfc.nasa.gov /planetary/factsheet/mercuryfact.html.

His full ratio table was as follows: Johannes Kepler, *Harmonies of the World*, trans. Charles Glenn Wallis (Annapolis, MD: St. John's Bookstore, 1939), 21.

Chandra recorded "sound" waves: Dana Bolles, "Black Hole Sound Waves," NASA Science, September 9, 2003, https://science.nasa.gov /science-news/science-at-nasa/2003/09sep_blackholesounds.

gold-plated gramophone discs were stored on board: "The Golden Record," NASA Jet Propulsion Laboratory, accessed January 14, 2022, https://voyager.jpl.nasa.gov/golden-record/.

waves of plasma were measured around 300 Hz: "Voyager Captures Sounds of Interstellar Space," NASA Jet Propulsion Laboratory, September 6, 2013, YouTube video, 1:01, https://www.youtube .com/watch?v=LIAZWb9_si4.

a few minutes short of twenty-four hours: Karl G. Jansky, "Electrical
    Disturbances Apparently of Extraterrestrial Origin," *Proceedings
    of the Institute of Radio Engineers* 21, no. 10 (October 1933):
    1387–98, https://doi.org/10.1109/JRPROC.1933.227458.

electric and magnetic fields in our atmosphere: "What Is a Schumann
    Resonance?," NASA Goddard Space Flight Center, accessed
    January 15, 2022, https://image.gsfc.nasa.gov/poetry/ask/q7
    68.html.

the unmanned *InSight* spacecraft: "MARS InSight Mission," NASA
    Science Mission Directorate, accessed January 15, 2022,
    https://mars.nasa.gov/insight/.

why a seismometer is so important: Jim Green, "Gravity Assist: Mars and
    InSight with Bruce Banerdt," NASA, May 2, 2018, https://www
    .nasa.gov/mediacast/gravity-assist-mars-and-insight-with
    -bruce-banerdt.

InSight recorded its first Marsquake: "First Likely Marsquake Heard
    by NASA's InSight," NASA Jet Propulsion Laboratory, April
    23, 2019, YouTube video, 1:09, https://www.youtube.com
    /watch?v=DLBP-5KoSCc.

Apollo 13 catapulted the third stage: "Apollo 13's Booster Impact,"
    NASA, March 23, 2010, https://www.nasa.gov/mission_pages
    /LRO/multimedia/lroimages/lroc-20100322-apollo13booster.html.

recent aurora australis events: Duane W. Hamacher, "Aurorae in
    Australian Aboriginal Traditions," *Journal of Astronomical
    History and Heritage* 16, no. 2 (2013): 211, https://arxiv.org
    /abs/1309.3367.

explanations of the aurora borealis: E. W. Hawkes, *The Labrador Eskimo*
    (Ottawa: Government Printing Bureau, 1916), 153.

departed souls that appears as the aurora borealis: Knud Rasmussen,
    Intellectual Culture of the Iglulik Eskimos (Copenhagen:
    Gyldendalske Boghandel, 1929), 95.

an inversion layer in the atmosphere: Unto K. Laine, "Noises as Well as Claps Associated with the Northern Lights," Aalto University, July 26, 2017, https://www.aalto.fi/en/news /noises-as-well-as-claps-associated-with-the-northern-lights.

B♭—Pitch Perfect

"the occasion...was de'void": "From the Archives 1919: The Treaty of Versailles signed", June 28, 2019, https://www.smh.com.au /world/europe/from-the-archives-1919-treaty-of-versailles -signed-20190619-p51zas.html

the war over standardized musical pitch: Fanny Gribenski, "Negotiating the Pitch: For a Diplomatic History of A, at the Crossroads of Politics, Music, Science and Industry," in *International Relations, Music and Diplomacy: Sounds and Voices on the International Stage*, ed. Frédéric Ramel and Cécile Prévost-Thomas (Cham, Switzerland: Palgrave Macmillan, 2018), 173–92, https://doi .org/10.1007/978-3-319-63163-9_8.

a conference was held in London: James A. R. Nafziger, Robert Kirkwood Paterson, and Alison Dundes Renteln, *Cultural Law: International, Comparative, and Indigenous* (Cambridge, UK: Cambridge University Press, 2014), 95.

"I added a third at the top": interview with Jim Reekes, April 26, 2021.

"ever after unable to perform": R. C. Bickerton and G. S. Barr, "The Origin of the Tuning Fork," *Journal of the Royal Society of Medicine* 80, no. 12 (December 1987): 771, https://doi .org/10.1177/014107688708001215.

Harold Rhodes: Gerald McCauley and Benjamin Bove, *Down the Rhodes: The Fender Rhodes Story* (Lanham, MD: Rowman & Littlefield/Hal Leonard, 2013).

Rhodes was issued with U.S. Patent 2,972,922: Harold B. Rhodes,

Electrical musical instrument in the nature of a piano, U.S. Patent 2,972,922, filed March 9, 1959, and issued February 28, 1961, https://patents.google.com/patent/US2972922.

In his 1683 sounding wheel experiment: Benjamin Wardhaugh, *Music, Experiment and Mathematics in England, 1653–1705* (Aldershot, UK: Ashgate, 2008), 107.

long and bitter feud with Sir Isaac Newton: Felicity Henderson, "Hooke, Newton, and the 'missing' portrait," Royal Society, December 3, 2010, https://royalsociety.org/blog/2010/12/hooke-newton-and-the-missing-portrait/.

B—Bridges over Troubled Waters

"I fled the building": "Hundreds evacuate shaking shopping mall in Seoul", *Korea Times*, July 5, 2011, https://www.koreatimes.co.kr/www/nation/2021/09/113_90253.html?KK.

"The fire station dispatched": Lee Hyo-sik, "Hundreds evacuate Techno-Mart after jolt", Korea Times, July 5, 2011, http://www.koreatimes.co.kr/www/news/nation/2011/07/117_90308.html.

"I put the machine": Allan L. Benson, "Nikola Tesla, Dreamer", *World To-day*, February 1, 1912.

a frightening chain reaction: Lan Chung, Taewon Park, and Sung Sik Woo, "Vertical Shaking Accident and Cause Investigation of 39-story Office Building," *Journal of Asian Architecture and Building Engineering* 15, no. 3 (September 2016): 619–25, https://doi.org/10.3130/jaabe.15.619.

hybrid mass damper: "Gangbyun Technomart," TESolution, accessed January 15, 2022, http://www.tesolution.com/technomart.html.

production of alternating current (AC): "Tesla Earthquake Machine [3D Printed]," Integza, September 29, 2019, YouTube video, 13:30, https://www.youtube.com/watch?v=l5rEW9QgJDc.

human perception of seismic sound: Patrizia Tosi, Paola Sbarra, and Valerio De Rubeis, "Earthquake Sound Perception," *Geophysical Research Letters* 39, no. 24 (December 19, 2012), https://doi.org/10.1029/2012GL054382.

On Boxing Day morning: "My Life Was Saved By An Elephant—Amber Owen," This Morning, June 8, 2016, YouTube video, 6:51, https://www.youtube.com/watch?v=BqHe2DEI1B4.

traveling away from the epicenter at almost 1,000 kilometers per hour: Eric Geist, "Tsunami Generation from the 2004 M=9.1 Sumatra-Andaman Earthquake," Pacific Coastal and Marine Science Center, October 8, 2018, https://www.usgs.gov /centers/pcmsc/science/tsunami-generation-2004-m91-sumatra -andaman-earthquake?qt-science_center_objects=0#qt-science _center_objects.

substantial number of reports of snakes: Kelin Wang et al., "Predicting the 1975 Haicheng Earthquake," *Bulletin of the Seismological Society of America* 96, no. 3 (July 2006): 757–95, https://doi .org/10.1785/0120050191.

producing infrasonic vocal rumbles: Michael Garstang, "Long-Distance, Low-Frequency Elephant Communication," *Journal of Comparative Physiology A* 190, no. 10 (November 2004): 791–805, https://doi.org/10.1007/s00359-004-0553-0.

elephants triangulate signals using their feet and ears: "When It Comes to Elephant Love Calls, the Answer Lies in a Boneshaking Triangle," Science Daily, February 14, 2009, https://www .sciencedaily.com/releases/2009/02/090213161038.htm.

two satellite-tracked collared Asian elephants: Eric Wikramanayake, Prithiviraj Fernando, and Peter Leimgruber, "Behavioral Response of Satellite-Collared Elephants to the Tsunami in Southern Sri Lanka," *Biotropica* 38, no. 6 (November 2006): 775–77, https://doi.org/10.1111/j.1744-7429.2006.00199.x.

detect a range of abiotic sounds: Michael Garstang and Michael C. Kelley, "Understanding Animal Detection of Precursor Earthquake Sounds," *Animals* 7, no. 9 (August 2017): 66, https://doi.org/10.3390/ani7090066.

Whitetail arrived back at his loft: Robert Krulwich, "After Tens of Thousands of Pigeons Vanish, One Comes Back," *National Geographic*, May 26, 2016, https://www.nationalgeographic .com/science/article/after-tens-of-thousands-of-pigeons-vanish -one-comes-back.

"I was absolutely amazed": "Racing Pigeon Returns—Five Years Late," *Manchester Evening News*, May 7, 2005, https://www .manchestereveningnews.co.uk/news/local-news/racing-pigeon -returns---five-1180418.

pointed the finger of blame at...the Concorde: Jean Varnier, Géraldine Ménéxiadis, and Ingrid Le Griffon, "Sonic Boom and Infrasound Emission from Concorde Airliner," *Proceedings of the Acoustics 2012 Nantes Conference* (April 2012), https://hal .archives-ouvertes.fr/hal-00811211/document.

pigeons can pinpoint their location: "Jonathan Hagstrum: Avian Navigation, Pigeon Homing and Infrasound," Sonic Acts, June 27, 2017, YouTube video, 35:36, https://www.youtube.com /watch?v=ALBOyHMuIgc.

generated by ocean storms: John A. Orcutt, "Quantitative Prediction of Seafloor Noise at Low Frequencies," in *Geology and Geophysics Program Summary for FY92*, ed. J. H. Kravitz (Arlington, VA: U.S. Navy Office of Naval Research, 1993), 208–10.

an ambient seismic wave field: Kiwamu Nishida, "Ambient Seismic Wave Field," *Proceedings of the Japan Academy, Series B Physical and Biological Sciences* 93, no. 7 (August 2017): 423–48, https:// doi.org/10.2183/pjab.93.026.

But not everyone loved the Concorde: "Concorde (Sonic Boom)," 946

Parl. Deb. H.C., March 23, 1978, cols. 1773–88, https://api .parliament.uk/historic-hansard/commons/1978/mar/23 /concorde-sonic-boom.

As a flying object travels: "Sonic Boom Explained—How It Is Created— Animated Graphics," Educational Video Library, April 1 2017, YouTube video, 5:54, https://www.youtube.com/watch ?v=1pf-Is2S1_Q.

snapping stay chains and bolts: *The Penny Cyclopaedia of the Society for the Diffusion of Useful Knowledge* (London: Charles Knight, 1842), 23:339.

London's Albert Bridge: "Albert Bridge (1950)," British Pathé, April 13, 2014, YouTube video, 0:50, https://www.youtube.com /watch?v=H8H_hSZ__-w.

Tacoma Narrows Bridge collapse of 1940: "Tacoma Bridge," Simon Lespérance, September 23, 2006, YouTube video, 4:12, https://www.youtube.com/watch?v=3mclp9QmCGs.

a very lucky escape: Howard Clifford, "Black and Blue and Lucky to Be Alive, the Last Person to Escape Galloping Gertie Tells His Story," *Tacoma News Tribune*, October 31, 2015, https://www .thenewstribune.com/news/local/article41608875.html.

The "blade of light": "Engineer Chris Wise Talks About the Millennium Bridge," ExpeditionWorkshed, March 20, 2013, YouTube video, 6:38, https://www.youtube.com/watch?v=SQEAj29IkNU.

subsequent tests revealed: Tony Fitzpatrick and Roger Ridsdill Smith, "Stabilising the London Millennium Bridge," *Ingenia* 9 (August 2001), https://www.ingenia.org.uk/ingenia/issue-9/stabilising -the-london-millennium-bridge.

C—Ghost Notes

Tandy's lab at Coventry University: Vic Tandy and Tony R. Lawrence, "The Ghost in the Machine," *Journal of the Society for Psychical Research* 62, no. 851 (April 1998).

In a 1976 report: Daniel Johnson, "Infrasound, Its Sources and Its Effects on Man," Aerospace Medical Research Laboratory, May 1, 1976, https://apps.dtic.mil/sti/pdfs/ADA032401.pdf.

notably in 1965 on Gemini 5: Curtis E. Larsen, "NASA Experience with Pogo in Human Spaceflight Vehicles," NASA Johnson Space Center, May 5, 2008, https://ntrs.nasa.gov/citations/20080018689.

some recent NASA research: Jonathan B. Clark, "Human Issues related to Spacecraft Vibration during Ascent," Consultant Report to the Constellation Program Standing Review Board.

"that there isn't anything.": Chris Arnot, "Ghost buster," *Guardian*, July 11, 2000, https://www.theguardian.com/education/2000/jul/11/highereducation.chrisarnot.

Researchers from Reading University: Aaron Watson and Dave Keating, "Architecture and Sound: An Acoustic Analysis of Megalithic Monuments in Prehistoric Britain," *Antiquity* 73, no. 280 (June 1999): 352–56, https://doi.org/10.1017/S0003598X00088281.

painted red dots: Iegor Reznikoff, "The Evidence of the Use of Sound Resonance from Palaeolithic to Medieval Times," in *Archeoacoustics*, ed. Chris Scarre and Graeme Lawson (Cambridge, UK: McDonald Institute for Archaeological Research, 2006), 77–84.

In 2008, a team of psychiatrists: Ian A. Cook, Sarah K. Pajot, and Andrew F. Leuchter, "Ancient Architectural Acoustic Resonance Patterns and Regional Brain Activity," *Time and Mind: The Journal of Archaeology, Consciousness and Culture* 1, no. 1 (March 2008): 95–104, https://doi.org/10.2752/175169608783489099.

didgeridoo and a small whistle pipe: "Tuvan Throat Singing," FNscarH, March 21, 2008, YouTube video, 1:43, https://www.youtube. com/watch?v=VTCJ5hedcVAh.

her piece for piano and electronics: "Infrasonic—Haunted Music?," Sarah Angliss, accessed January 15, 2022, https://www.sara hangliss.com/infrasonic/.

the UK's Royal Naval Physiological Laboratory: Edward Cudahy and Stephen Parvin, "The Effects of Underwater Blast on Divers," Naval Submarine Medical Research Laboratory, February 8, 2001, https://apps.dtic.mil/sti/pdfs/ADA404719.pdf.

Db—Fix in the Mix

a pair of handcuffs and a wooden rattle: David Cross, "On the Beat in Birmingham," BBC History, February 17, 2011, https://www .bbc.co.uk/history/british/victorians/beat_01.shtml.

the two-tone whistle: Chris Heather, "The Metropolitan Police: An Introduction to Records of Service 1829–1958," National Archives, June 10, 2011, https://media.nationalarchives.gov .uk/index.php/the-metropolitan-police-an-introduction-to -records-of-service-1829-1958-2/.

Joseph and James Hudson: "The First Whistle," ACME Whistles, accessed January 15, 2022, https://www.acmewhistles.co.uk /the-first-whistle.

a two-note prototype whistle: "The Metropolitan Police Whistle," Birmingham Heritage Walking Tours, June 26, 2019, YouTube video, 0:46, https://www.youtube.com/watch?v=1JILRoNXHMo.

The infantry training manual of 1914: "War Whistles History—Whistles Used in WWI by British, German and Commonwealth Forces," War Whistles, accessed January 15, 2022, https://www.war whistles.com/ww1-whistles-history.html.

bilabial plosives: Michael Ashby and John Maidment, *Introducing Phonetic Science* (Cambridge, UK: Cambridge University Press, 2005), 183.

problematic vocal sounds: Ian Maddieson, "Voicing and Gaps in Plosive Systems," World Atlas of Language Structures Online, accessed January 15, 2022, https://wals.info/chapter/5.

Paul Lueg in Germany in the 1930s: Paul Lueg, Process of silencing sound oscillations, U.S. Patent 2,043,416, filed March 8, 1934, and issued June 9, 1936, https://patents.google.com/patent/US2043416A/en.

The *Los Angeles Times* reported: Harry Nelson, "Concern Rises for Voyager Pilots' Health: Noise in Tiny Cabin Could Cause Partial Loss of Hearing," *Los Angeles Times*, December 19, 1986, https://www.latimes.com/archives/la-xpm-1986-12-19-mn-3538-story.html.

Dr. Henning von Gierke: Paul Schomer, "Dr. Henning von Gierke—My Mentor, Everyone's Mentor," *Journal of the Acoustical Society of America* 123, no. 5 (June 2008): 3132, https://doi.org/10.1121/1.2933086.

principle of active noise cancellation: Eric Schulze, "Ask Smithsonian: How Do Noise-Canceling Headphones Work?," *Smithsonian Magazine*, accessed January 15, 2022, https://www.smithsonianmag.com/videos/ask-smithsonian-how-do-noise-canceling-headph/.

From the online demonstrations: "Force SV: The Wave Extinguisher," Force SV, accessed January 15, 2022, https://www.forcesv.com.

skeptical of most of it: Dom Monks interview, October 12, 2020

D—Only the Lonely

highly classified SOSUS: "Origins of SOSUS," Submarine Force Pacific, accessed January 15, 2022, https://www.csp.navy.mil/cus /About-IUSS/Origins-of-SOSUS/.

deep sound channel or SOFAR: "What is SOFAR?," National Ocean Service, updated February 26, 2021, https://oceanservice .noaa.gov/facts/sofar.html.

One man was especially important: "William A. Watkins," Woods Hole Oceanographic Institution, accessed January 15, 2022, https:// www.whoi.edu/who-we-are/about-us/people/obituary/william -a-watkins/.

SOSUS started recording a distinctive whale call: William A. Watkins et al., "Twelve Years of Tracking 52-Hz Whale Calls from a Unique Source in the North Pacific," *Deep Sea Research Part 1: Oceanographic Research Papers* 51, no. 12 (December 2004): 1889–901, https://doi.org/10.1016/j.dsr.2004.08.006.

But then it started wandering: Lonny Lippsett, "A Lone Voice Crying in the Watery Wilderness," *Oceanus*, April 5, 2005, https:// www.whoi.edu/oceanus/feature/a-lone-voice-crying-in-the -watery-wilderness.

fin and blue whale hybrid: Martine Bérubé and Alex Aguilar, "A New Hybrid between a Blue Whale, *Balaenoptera musculus*, and a Fin Whale, *B. physalus*: Frequency and Implications of Hybridization," *Marine Mammal Science* 14, no. 1 (January 1998): 82–98, https://doi.org/10.1111/j.1748-7692.1998.tb00692.x.

an increasingly noisy ocean: Peter Lloyd Tyack and Vincent M. Janik, "Effects of Noise on Acoustic Signal Production in Marine Mammals," in *Animal Communication and Noise*, ed. Henrik Brumm (Berlin, Germany: Springer-Verlag, 2013), 251–71, https://doi.org/10.1007/978-3-642-41494-7_9.

baleen whales reduced their calls: Mariana L Melcón et al., "Blue

Whales Respond to Anthropogenic Noise," *PLOS ONE* 7, no. 2 (February 2012): e32681, https://doi.org/10.1371/journal .pone.0032681.

Studies of blackbirds: Erwin Nemeth and Henrik Brumm, "Blackbirds Sing Higher-Pitched Songs in Cities: Adaptation to Habitat Acoustics or Side-Effect of Urbanization?," *Animal Behavior* 28, no. 3 (September 2009): 637–41, https://doi.org/10.1016/j .anbehav.2009.06.016.

web-building European garden spider: Chung-Huey Wu and Damian O Elias, "Vibratory Noise in Anthropogenic Habitats and Its Effect on Prey Detection in a Web-Building Spider," *Animal Behavior* 90, (April 2014): 47–56, https://doi.org/10.1016/j .anbehav.2014.01.006.

Phidippus audax: Paul S. Shamble et al., "Airborne Acoustic Perception by a Jumping Spider," *Current Biology* 26, no. 21 (November 2016): 2913–20, https://doi.org/10.1016/j.cub.2016.08.041.

Male peacock spiders tap their feet: Damian O. Elias et al., "Seismic Signals in a Courting Male Jumping Spider," *Journal of Experimental Biology* 206, no. 22 (November 2003): 4029–39, https://doi.org/10.1242/jeb.00634.

Through the power of dance and drumming: "Peacock Spiders, Dance for Your Life!," BBC, November 21, 2019, YouTube video, 4:48, https://www.youtube.com/watch?v=5qkzwG2lLPc.

Strings of silk can be tuned by spiders: Beth Mortimer et al., "The Speed of Sound in Silk: Linking Material Performance to Biological Function," *Advanced Materials* 26, no. 30 (June 2014): 5179–83, https://doi.org/10.1002/adma.201401027.

the population of invertebrates has fallen: Rodolfo Dirzo et al., "Defaunation in the Anthropocene," *Science* 345, no. 6195 (July 2014): 401–6, https://doi.org/:10.1126/science.1251817.

four hundred million tons of food each year: Martin Nyffeler and

Klaus Birkhofer, "An Estimated 400–800 Million Tons of Prey Are Annually Killed by the Global Spider Community," *Die Naturwissenschaften* 104, no. 3–4 (March 2017): 30, https://doi.org/10.1007/s00114-017-1440-1.

this form of wave pollution: S. Shepherd et al., "Extremely Low Frequency Electromagnetic Fields Impair the Cognitive and Motor Abilities of Honey Bees," *Scientific Reports* 8, no. 1 (May 2018): 7932, https://doi.org/10.1038/s41598-018-26185-y.

a bee's wing frequency is 256 Hz: Yuji Hasegawa and Hidetoshi Ikeno, "How Do Honeybees Attract Nestmates Using Waggle Dances in Dark and Noisy Hives?," *PLOS ONE* 6, no. 5 (May 2011): e19619, https://doi.org/10.1371/journal.pone.0019619.

the resonant frequencies of rats' individual whiskers: Maria A. Neimark et al., "Vibrissa Resonance as a Transduction Mechanism for Tactile Encoding," *Journal of Neuroscience* 23, no. 16 (July 2003): 6499–509, https://doi.org/10.1523/JNEUROSCI.23-16-06499.2003.

a grid of rats' whiskers: S. B. Vincent, "The Tactile Hair of the White Rat," *Journal of Comparative Neurology* 23, no. 1 (February 1913): 1–34, https://doi.org/10.1002/cne.900230101.

direction of the wind: Yan S. W. Yu et al., "Whiskers Aid Anemotaxis in Rats," *Science Advances* 2, no. 8 (August 2016): e1600716, https://doi.org/10.1126/sciadv.1600716.

plants' aural and musical abilities: Richard M. Klein and Pamela C. Edsall, "On the Reported Effects of Sound on the Growth of Plants," *BioScience* 15, no. 2 (February 1965): 125–26, https://doi.org/10.2307/1293353.

impatiens as their subjects: Margaret E. Collins and John E. K. Foreman, "The Effect of Sound on the Growth of Plants," *Canadian Acoustics* 29, no. 2 (2001): 3–8, https://jcaa.caa-aca.ca/index.php/jcaa/article/view/1358.

bioacoustician Monica Gagliano: Monica Gagliano, Stefano Mancuso, and Daniel Robert, "Towards Understanding Plant Bioacoustics," *Trends in Plant Science* 17, no. 6 (June 2012): 323–25, https://doi.org/10.1016/j.tplants.2012.03.002.

I quizzed Joshua Zeman: Joshua Zeman interview, January 12, 2021

E♭—Of Sound Mind

one hundred billion neurons: Priyanka A. Abhang, Bharti W. Gawali, and Suresh Mehrotra, "Technological Basics of EEG Recording and Operation of Apparatus," in *Introduction to EEG- and Speech-Based Emotion Recognition* (London: Academic Press, 2016), 19–50, https://doi.org/10.1016/B978 -0-12-804490-2.00002-6.

branches called dendrites: Alan Woodruff, "What is a Neuron?," Queensland Brain Institute, updated August 13, 2019, https:// qbi.uq.edu.au/brain/brain-anatomy/what-neuron.

recent international research: Victoria Leong, Elizabeth Byrne, Kaili Clackson, Stanimira Georgieva, Sarah Lam, Sam Wass, "Speaker gaze increases information coupling between infant and adult brains," *Proceedings of National Academy of Sciences* 114, no. 50 (November 2017), 13290-13295, https://www.pnas. org/doi/10.1073/pnas.1702493114.

shared attention in a classroom: Suzanne Dikker, Lu Wan, Ido Davidesco, Lisa Kaggen, Matthias Oostrik, James McClintock, Jess Rowland, Georgios Michalareas, Jay J. Van Bavel, Mingzhou Ding, David Poeppel, "Brain-to-Brain Synchrony Tracks Real-World Dynamic Group Interactions in the Classroom," *Current Biology* 27, no. 9, (May 2017), 1375-1380, https://www.science-direct.com/science/article/pii/S0960982217304116.

Ross reported: Alex Ross, "Listening to Prozac...Er, Mozart,"

New York Times, August 28, 1994, https://www.nytimes.com/1994/08/28/arts/classical-view-listening-to-prozac-er-mozart.html.

Center for the Neurobiology of Learning and Memory: Frances H. Rauscher, Gordon L. Shaw, and Catherine N. Ky, "Music and Spatial Task Performance," *Nature* 365, no. 6447 (October 1993): 611, https://doi.org/10.1038/365611a0.

"So, does the Mozart effect exist?": J. S. Jenkins, "The Mozart effect," *Journal of the Royal Society of Medicine* 94, no. 4 (April 2001), 170-172, https://www.ncbi.nlm.nih.gov/pmc/articles/PMC1281386/.

Another piece of research: John R. Hughes and John J. Fino, "The Mozart Effect: Distinctive Aspects of the Music—A Clue to Brain Coding?," *Clinical EEG and Neuroscience* 31, no. 2 (April 2000): 94–103, https://doi.org/10.1177/155005940003100208.

A report by ABC News in 2009: Rebecca Lee, "The Moozart Effect," ABC News, January 8, 2009, https://abcnews.go.com/Technology/story?id=3213324&page=1.

Leaving aside the claim: Clare G. Harvey, *The Practitioner's Encyclopedia of Flower Remedies: The Definitive Guide to All Flower Essences, Their Making and Uses* (London: Jessica Kingsley Publishers, 2014), 455.

stories have attracted academic interest: "Levitation of Frogs and Pyramids?," Bruce Drinkwater, December 1, 2016, https://brucedrinkwater.com/2016/12/01/acoustic-levitation-of-the-pyramids-is-theoretically-possible/.

E—Somewhere over the Rainbow

first durable color photograph was taken: "A Short History of Colour Photography," The National Science and Media Museum,

July 7, 2020, https://www.scienceandmediamuseum.org.uk /objects-and-stories/history-colour-photography.

illustration in *Opticks*: Isaac Newton, *Opticks: or, A Treatise of the Reflexions, Refractions, Inflexions and Colours of Light* (London, 1704), 186, https://library.si.edu/digital-library/book/opticks treatise00newta.

little tartan rosette: "First Colour Photographic Image by Maxwell," James Clerk Maxwell Foundation, accessed January 16, 2022, https://clerkmaxwellfoundation.org/html/first_color_photo graphic_image.html.

tingling on the lips and tongue: Nobuhiro Hagura, Harry Barber, and Patrick Haggard, "Food Vibrations: Asian Spice Sets Lips Trembling," *Proceedings of the Royal Society B* 280, no. 1770 (November 2013): 1680, https://doi.org/10.1098 /rspb.2013.1680.

"transverse myelitis, and encephalitis": "Paresthesia Information Page," National Institute of Neurological Disorders and Stroke, https://www.ninds.nih.gov/Disorders/All-Disorders /Paresthesia-Information-Page.

dense in the clitoris: Bruno Grignon, review of *Anatomic Study of the Clitoris and the Bulbo-Clitoral Organ*, by Vincent Di Marino and Hubert Lepidi, *Surgical and Radiologic Anatomy* 37 (May 2015): 1291, https://doi.org/10.1007/s00276-015-1490-z.

sex historian Hallie Lieberman: Hallie Lieberman, "Selling Sex Toys: Marketing and the Meaning of Vibrators in Early Twentieth-Century America," *Enterprise and Society* 17, no. 2 (June 2016): 393–433, https://doi.org/10.1017/eso.2015.97.

explore an antique vibrator: "Unboxing 1930s Women's Products," mycallaly, May 20, 2020, YouTube video, 1:00, https://www .youtube.com/watch?v=_KNIbxL-G7Y.

quite the opposite: Umit Sayin et al., "Effects of Low Frequency Bullet

Vibrators on the Clitoral Orgasm and Sexual Lives of Turkish Women," *Journal of Sexual Medicine* 12 (May 2015): 217, https://www.researchgate.net/publication/292817627_EFFECTS _OF_LOW_FREQUENCY_BULLET_VIBRATORS_ON _THE_CLITORAL_ORGASM_AND_SEXUAL_LIVES _OF_TURKISH_WOMEN.

smell is driven chemically: Simon Cotton, "If It Smells—It's Chemistry," Royal Society of Chemistry, March 1, 2009, https://edu.rsc .org/feature/if-it-smells-its-chemistry/2020168.article.

vibrational theory of olfaction: Simon Gane et al., "Molecular Vibration-Sensing Component in Human Olfaction," *PLOS ONE* 8, no. 5 (January 2013): e55780, https://doi.org/10.1371/journal .pone.0055780.

our noses "listen" to vibrations: "The Science of Scent," Luca Turin, February 2005, TED video, 15:40, https://www.ted.com /talks/luca_turin_the_science_of_scent?language=en.

all molecules vibrate: Andrea Rinaldi, "Do Vibrating Molecules Give Us Our Sense of Smell?," *Science*, February 14, 2011, https://www.sciencemag.org/news/2011/02/do-vibrating-molecules -give-us-our-sense-smell.

BBC documentary on this subject: Jim Al-Khalili, "The Secrets of Quantum Physics," BBC, December 2014, https://www.bbc .co.uk/programmes/b04v85cj.

glamorous of smelly obsessions: Sam Lear, "Perfumery: The Molecular Art Form," *Chemistry World*, October 2, 2015, https://www .chemistryworld.com/features/perfumery-the-molecular-art -form/9003.article.

smell might remind us of another person: Jordan Gaines Lewis, "Smells Ring Bells: How Smell Triggers Memories and Emotions," *Psychology Today*, January 12, 2015, https://www.psychologytoday.com/gb/blog/brain-babble/20

1501/smells-ring-bells-how-smell-triggers-memories-and
-emotions.

compounds called esters: Brian C. Smith, "The C=O Bond, Part VI:
Esters and the Rule of Three," *Spectroscopy* 33, no. 7 (July
2018): 20–23, https://www.spectroscopyonline.com/view
/co-bond-part-vi-esters-and-rule-three.

notable study in 1962: S. Rosen et al., "Presbycusis Study of a Relatively
Noise Free Population of the Sudan," *Annals of Otology,
Rhinology and Laryngology* 71 (September 1962): 135-152.

Shepard started suffering: Richard Menger et al., "Rear Admiral
(Astronaut) Alan Shepard: Ménière's Disease and the Race to
the Moon," *Journal of Neurosurgery* 131, no. 1 (February 2019):
304–10, https://doi.org/10.3171/2018.9.JNS182522.

world's first true cochlear implant: Albert Mudry and Mara Mills, "The
Early History of the Cochlear Implant: A Retrospective," *JAMA
Otolaryngology Head Neck Surgery* 139, no. 5 (May 2013): 446–
53, https://doi.org/10.1001/jamaoto.2013.293.

much controversy still surrounds: Caroline Praderio, "Why Some
People Turned Down a 'Medical Miracle' and Decided to
Stay Deaf," *Insider*, January 3, 2017, https://www.insider.com
/why-deaf-people-turn-down-cochlear-implants-2016-12.

F—Acoustic Ammunition

suddenly swerved violently to the left: "Stunning Claims and
Controversy at Royal Ascot in 1988," BBC Sport, June 16, 2011,
https://www.bbc.co.uk/sport/av/horse-racing/13799912.

"the ultrasonic device was the centerpiece": "Inventer [sic] of
Ultrasonic Gun Testifies at Race Fraud Trial," Associated
Press, November 3, 1989, https://apnews.com/article/fc6a3f
dee491536fa0d4e27286ab9640.

"I got on the horse": "Jockey Tells of 'Lethal Stun Gun,'" *Herald of Scotland*, November 7, 1989, https://www.heraldscotland.com /news/11967543.jockey-tells-of-lethal-stun-gun-tests/.

the process of presbycusis begins: "Take the High-Frequency Hearing Test," *National Geographic*, November 8, 2013, YouTube video, 2:50, https://www.youtube.com/watch?v=sZHW Y1KBHwc.

outside their local corner shop: "Noise Machine Deters Shop Gangs," BBC News, November 8, 2005, http://news.bbc.co.uk/1 /hi/wales/south_east/4415318.stm.

The first was from human rights activists: Sarah Campbell, "Now Crime Gadget Can Annoy Us All," BBC News, December 2, 2008, http://news.bbc.co.uk/1/hi/uk/7759818.stm.

"demonizing children and young people": Kate Martin, "Campaign targets 'Mosquito' noise pollution," edie newsroom, February 19, 2008, https://www.edie.net/news/1/Campaign-targets-Mosquito -noise-pollution/14228/.

"rather than to our kids": Rosalind Ryan, "Call for ban on audio device that targets young ears," *Guardian*, February 12, 2008, https:// www.theguardian.com/society/2008/feb/12/mosquito.young .people.

surprising level of manufacturing processes: Lavi Bigman, "Standing Waves in the Synagogue: The Physics of the Shofar," Davidson Institute of Science Education, September 8, 2018, https:// davidson.weizmann.ac.il/en/online/sciencepanorama/standing -waves-synagogue-physics-shofar.

Gershwin's orchestrator, Ferde Grofé: Charles Schwartz, *Gershwin: His Life and Music* (Indianapolis: Da Capo, 1979), 81–82.

firing at any possible threats: Stephen Robinson, "Bombed US Warship Was Defended by Sailors with Unloaded Guns," *Telegraph*, November 15, 2000, https://www.telegraph.co.uk/news

/worldnews/middleeast/yemen/1374316/Bombed-US-warship
-was-defended-by-sailors-with-unloaded-guns.html.

Seabourn Spirit cruise ship: "Seabourn Ship Used Sonic Defense to Ward
off Pirates," *Maritime Executive*, November 9, 2005, https://
www.maritime-executive.com/article/2005-11-09seabourn-ship
-used-sonic-defense-to-wa.

crew soon came under fire: "'I Beat Pirates with a Hose and Sonic
Cannon,'" BBC News, May 17, 2007, http://news.bbc.co.uk/1
/hi/uk/6664677.stm.

other orchestral musicians: "Professional Musicians Run Almost
Four-Fold Risk of Noise Induced Deafness," Science Daily,
April 30, 2014, https://www.sciencedaily.com/releases/2014
/04/140430192647.htm.

opponents of the LRAD: Ben Kesslen, "'Plug Your Ears and Run':
NYPD's Use of Sound Cannons Is Challenged in Federal
Court," NBC News, May 22, 2019, https://www.nbcnews
.com/news/us-news/plug-your-ears-run-nypd-s-use-sound
-cannons-challenged-n1008916.

"Some 1.1 billion teenagers": "1.1 billion people at risk of hearing loss",
World Health Organization, March 10, 2015, https://www
.who.int/vietnam/news/detail/10-03-2015-1.1-billion-people
-at-risk-of-hearing-loss.

nineteenth-century Dutch train spotters: Reto U. Schneider, *The Mad
Science Book: 100 Amazing Experiments from the History of
Science* (London: Penguin, 2008).

Buys Ballot, with a slightly tiresome: Buijs Ballot, "Akustische Versuche
auf der Niederländischen Eisenbahn, nebst gelegentlichen
Bemerkungen zur Theorie des Hrn. Prof. Doppler," *Annalen
der Physik* (1845): https://onlinelibrary.wiley.com/doi/10.1002
/andp.18451421102.

conductor Charles Hazlewood: "The Doppler Effect with Charles
    Hazlewood," BBC Radio 4, August 10, 2017, https://www.bbc
    .co.uk/programmes/p05bxzhh.

A wonderful 1950 photograph: "Denver Post Archives," Getty Images,
    accessed January 16, 2022, https://www.gettyimages.co.uk
    /detail/news-photo/samuel-bagno-of-new-city-inventor-of-the
    -burglar-news-photo/837925420.

journal *American Scientist*: Eugene L. Fuss, "The Technology of Burglar
    Alarm Systems," *American Scientist* 72, no. 4 (1984): 334–37,
    http://www.jstor.org/stable/27852756.

celeb picture of 1973: "Andy Warhol Carrying a Polaroid SX-70
    Camera," Flickr, June 9, 2010, https://www.flickr.com/photos
    /superernie/4685835141.

history and physics of the SX-70: Dan MacIsaac and Ari Hämäläinen,
    "Physics and Technical Characteristics of Ultrasonic Systems,"
    Physics Teacher 40, no. 1 (January 2002): 39–46, https://doi
    .org/10.1119/1.1457828.

G♭—Sleight of Ear

quality of cucumbers: "Laying Down Quality Standards for Cucumbers,"
    European Commission Regulation No. 1677/88, June 15, 1988,
    https://op.europa.eu/en/publication-detail/-/publication
    /e376f15f-007d-40ba-90a4-fa7f6f97341f/language-en.

BBC News article of 2008: Finlo Rohrer, "Will We Eat Wonky Fruit
    and Veg?," BBC News Magazine, November 12, 2008, http://
    news.bbc.co.uk/1/hi/magazine/7724347.stm.

computer-led pitch correction: Andy Hildebrand, interview, NAMM
    Oral History Program, January 18, 2012, https://www.namm
    .org/library/oral-history/andy-hildebrand.

Hildebrand once said of this work: "Innovative Lives: Andy Hildebrand,"

Lemelson Center, April 19, 2018, YouTube video, 1:01:37, https://www.youtube.com/watch?v=DYFVa1y6kGI.

In a 1999 interview: Sue Sillitoe, "Recording Cher's 'Believe'—Mark Taylor & Brian Rawling," *Sound on Sound*, February 1999, https://www.soundonsound.com/techniques/recording-cher -believe.

"The 50 Worst Inventions": Dan Fletcher, "The Worst 50 Inventions—No. 15, Auto-Tune," *Time*, May 27, 2010, http://content.time.com /time/specials/packages/article/0,28804,1991915_1991909 _1991903,00.html.

Hildebrand recently said: "Andy Hildebrand," interview, NAMM Oral History Program, January 18, 2012, https://www.namm.org /library/oral-history/andy-hildebrand.

have been his sketches: Roger N. Shepard, *Mind Sights: Original Visual Illusions, Ambiguities, and Other Anomalies, with a Commentary on the Play of Mind in Perception and Art* (New York: W. H. Freeman, 1990).

Shepard-Risset glissando: Eveline Vernooij et al., "Listening to the Shepard-Risset Glissando: The Relationship between Emotional Response, Disruption of Equilibrium, and Personality," *Frontiers in Psychology* 7 (March 4, 2016): 300, https://doi.org/10.3389 /fpsyg.2016.00300.

After setting up her research lab: "Biography," Diana Deutsch, accessed January 16, 2022, http://deutsch.ucsd.edu/psychology/pages .php?i=218.

sometimes behave so strangely: Diana Deutsch, *Musical Illusions and Phantom Words: How Music and Speech Unlock Mysteries of the Brain* (New York: Oxford University Press, 2019), 151.

into a *musical* phrase: "Musical Illusions Perfect Pitch and Other Curiosities with Diana Deutsch—To Be Musical," University of California Television, April 11, 2013, YouTube video,

53:58 (particularly 30:14), https://www.youtube.com/watch ?v=pBeDn8XHhKU&t=1827s.

G—High Times

United States' greatest engineers: Don Murray, "Percy Spencer and His Itch to Know," *Reader's Digest*, August 1958, 114, https://www .smecc.org/percy_spenser_biography.htm.

mechanized panpipes: R. Nave, "The Magnetron," Georgia State University, accessed January 16, 2022, http://hyperphysics .phy-astr.gsu.edu/hbase/Waves/magnetron.html.

surface of your teeth and gums: R. P. Howlin et al., "Removal of Dental Biofilms with an Ultrasonically Activated Water Stream," *Journal of Dental Research* 94, no. 9 (September 2015): 1303–9, https://doi.org/10.1177/0022034515589284.

teeth every two hours: "Does Barry Manilow Really Brush His Teeth Every Two Hours?," *The Michael Ball Show*, BBC Sounds, accessed January 16, 2022, https://www.bbc.co.uk/sounds /play/p03xx83s.

developed echocardiography in 1953: Joseph Woo, "A Short History of the Development of Ultrasound in Obstetrics and Gynecology," Obstetric Ultrasound History Web, 1998, https://www.ob -ultrasound.net/history1.html.

continuous ink jet recorder patent: C. H. Hertz and Sven-Inge Simonsson, Ink jet recorder, U.S. Patent 3,416,153, filed October 6, 1966, and issued December 10, 1968, https:// patents.google.com/patent/US3416153A/en.

solar panels onto the surface of soap bubbles: Nisha Gaind, "The Best Science Images of 2020," *Nature*, December 14, 2020, https:// www.nature.com/immersive/d41586-020-03436-5/index.html.

Messing with our atoms: "How Radiation Harms Cells," Radiation Effects

Research Foundation, accessed January 16, 2022, https://www.rerf.or.jp/en/about_radiation/how_radiation_harms_cells_e/.

Bikini Atoll on that morning: "Bikini Atoll Evacuation Prior to Crossroads Atomic Bomb Test 70594," Periscope Film, October 8, 2016, YouTube video, 10:39, https://www.youtube.com/watch?v=zri2knpOSqo.

fisherman Ōishi Matashichi described it as: Ōishi Matashichi, The Day the Sun Rose in the West: Bikini, the Lucky Dragon, and I (Honolulu: University of Hawai'i Press, 2011).

radioactive isotopes: Mark Schreiber, "Lucky Dragon's Lethal Catch," Japan Times, March 18, 2012, https://www.japantimes.co.jp/life/2012/03/18/general/lucky-dragons-lethal-catch/.

skin burns, bleeding gums, and hair loss: "Daigo Fukuryu Maru Survivor Account," Nippon TV News 24 Japan, August 16, 2019, YouTube video, 2:24, https://www.youtube.com/watch?v=t_p_WBJmbvk.

"the wind failed to follow the predictions" and "present threat of these radioactive frequencies": "This Concrete Dome Holds a Leaking Toxic Timebomb," ABC News In-depth, November 21, 2017, YouTube video, 41:34, https://www.youtube.com/watch?v=autMHvj3exA.

In a recent investigative article: Maveric K. I. L. Abella et al., "Background Gamma Radiation and Soil Activity Measurements in the Northern Marshall Islands," Proceedings of the National Academy of Sciences of the United States of America 116, no. 31 (July 15, 2019): 15425–34, https://doi.org/10.1073/pnas.1903421116.

There are three types of radiation: "Ionizing Radiation, Health Effects and Protective Measures," World Health Organization, April 29, 2016, https://www.who.int/news-room/fact-sheets/detail/ionizing-radiation-health-effects-and-protective-measures.

reported from Chernobyl: Victoria Gill, "Chernobyl: The End of a Three-Decade Experiment," BBC News, February 14, 2019, https://www.bbc.co.uk/news/science-environment-47227767.

mathematical symbol X: H. J. W. Dam, "The New Marvel in Photography," *McClure's* 6, no. 5 (April 1896): 403–15.

the quartz crisis: Pierre-Yves Donzé, *A Business History of the Swatch Group* (Basingstoke, UK: Palgrave MacMillan, 2014).

Even as far back as 1879: Sir William Thomson & Peter Guthrie Tait, *Treatise on Natural Philosophy* (Cambridge, UK: Cambridge University Press, 1879).

which NASA says: "Time Conventions," NASA Solar System Explorations: Basics of Space Flight, accessed January 16, 2022, https://solarsystem.nasa.gov/basics/chapter2-3/.

"The collision occurred": "Awful Railroad Accident," *New York Times*, August 13, 1853, http://www.interment.net/data/train-wrecks /providence-and-worcester-railroad-collision-1853.htm.

millisecond for every six weeks of spaceflight: Arielle Samuelson, "NASA's Deep Space Atomic Clock Will Transform Space Exploration," SciTechDaily, June 19, 2019, https://scitechdaily .com/nasas-deep-space-atomic-clock-will-transform-space- exploration/.

MUSICAL FREQUENCY TABLE

NOTE	HERTZ	
F_6	1396.91	Infamous top soprano note (Mozart, "Queen of the Night")
C_6	1046.50	Last high trumpet note (Glenn Miller, "In the Mood")
B_5	987.77	First violin note at start (Bee Gees, "More Than a Woman")
Bb_5	923.33	First high trumpet note (John Williams, "Star Wars (Main Theme)")
A_5	880.00	Last note of classic Nokia ringtone (Francisco Tárrega, *Gran Vals*)
G_5	783.99	First violin note (Mozart, *Eine kleine Nachtmusik*)
F_5	689.46	First harmonica note (Beatles, "Love Me Do") First piano melody note (Mozart, *Elvira Madigan*)
E_5	659.25	First violin note (Vivaldi, "Spring") First saxophone note (George Michael, "Careless Whisper")
$D\#_5$	622.25	Last vocal note (Whitney Houston, "I Will Always Love You") First vocal note (Aretha Franklin, "Respect")
D_5	587.33	First soprano note (Handel, "Hallelujah Chorus")
$C\#_5$	554.37	First vocal note (Debbie Harry, "Heart of Glass")
C_5	523.25	First vocal note (Prince, "Kiss")

NOTE	HERTZ	
B_4	493.88	First solo guitar note (Eric Clapton, "Wonderful Tonight") First melody note (Brahms, "Lullaby")
Bb_4	466.16	First vocal note (Little Richard, "Good Golly, Miss Molly")
A_4	440	Oboe's note for orchestral tuning
$G\sharp_4$	415.3	First vocal note (Stevie Wonder, "Isn't She Lovely") First vocal note (Julie Covington, "Don't Cry for Me Argentina")
G_4	392.00	First violin note (Beethoven, Symphony no. 5) Opening note (*Close Encounters* "conversation")
F_4	349.23	First "La" (Kylie Minogue, "Can't Get You Out of My Head") First "wop bop a loo la" (Little Richard, "Tutti Frutti")
E_4	329.63	First "Oh" (Billy Joel, "Uptown Girl") First vocal note (Elvis Presley, "Heartbreak Hotel") First vocal note (Chubby Checker, "The Twist")
Eb_4	311.13	Last vocal note (Frank Sinatra, "New York, New York") First guitar note (Jimi Hendrix, "Voodoo Child")
D_4	293.66	First vocal note (Phil Collins, "In the Air Tonight") "Desmond" pitch (Beatles, "Ob-La-Di, Ob-La-Da") Opening phrase (Puccini, "Nessun dorma")
Db_4	277.18	First vocal note (Gene Kelly, "Singin' in the Rain") First vocal note (10cc, "I'm Not in Love")
C_4	261.63	MIDDLE C First trumpet note (Strauss, *Also sprach Zarathustra*) First vocal note (Idina Menzel, "Let It Go")
B_3	246.94	First vocal note (Ray Charles, "Georgia on My Mind")
Bb_3	233.08	First saxophone note (Dave Brubeck, "Take Five")
A_3	220.00	First vocal note (Elvis Presley, "Love Me Tender")
Ab_3	207.65	First saxophone note (Glenn Miller, "In the Mood") First vocal note (Andy Williams, "Moon River")
G_3	196.00	First violin melody note (Brahms, last movement, Symphony no. 1)
$F\sharp_3$	185.00	First vocal note (Beach Boys, "Barbara Ann")
E_3	164.81	First guitar note (Beatles, "Day Tripper")
Eb_3	155.56	First "Ah" (David Bowie, "Let's Dance")

NOTE	HERTZ	
D₃	146.83	First clarinet note for the cat (Prokofiev, *Peter and the Wolf*) First vocal note (Johnny Cash, "Ring of Fire")
D♭₃	138.59	First synth note (Vangelis, "Chariots of Fire")
C₃	130.81	First vocal note (Chet Baker, "My Funny Valentine") First guitar note (Survivor, "Eye of the Tiger")
B₂	123.47	First guitar note (Rolling Stones, "Satisfaction") First cello note (Schubert, *Unfinished* Symphony) First bass guitar note (Pink Floyd, "Money") First vocal note (Fats Domino, "Blueberry Hill")
B♭₂	116.54	First trombone note (Burt Bacharach, "What the World Needs Now")
G₂	98.00	First piano note (Erik Satie, "Gymnopédie no. 1")
C₂	65.41	First piano note (Dionne Warwick, "Do You Know the Way to San Jose")
A₂	55.00	First note of famous bass bit (Fleetwood Mac, "The Chain")
A♭₁	51.91	First bass note (Stravinsky, *The Firebird*)
E₁	41.20	First slap bass note (Frankie Goes to Hollywood, "Relax") First bass note (John Williams, "Main Title (Theme from *Jaws*)")
D₁	36.71	First piano note (Buddy Holly, "Look at Me")

ACKNOWLEDGMENTS

It has been a true privilege to meet so many fascinating, passionate, clever, and downright lovely people through my journey of writing *What the Ear Hears (and Doesn't)*. For the past six years, my professional life has expanded beyond narrow conversations with musicians and media people, introducing me to an amazing world of physicists, chemists, neurologists, surgeons, perfumers, dentists, psychologists, biologists, archaeoacousticians...

First, I must thank those who have given freely of their time. Thank you to all those who have allowed me to scratch the surface of their deep knowledge, their passion, and lifelong work—I apologize for asking, at times, the dumbest of questions.

George Bevan; Richard Canter (visiting professor of surgical education, Nuffield Department of Surgical Science, University of Oxford); Stephen Cooper; Redvers Daborn; Diana Deutsch (professor of psychology, University of California, San Diego); Paul Devereux; Chris Domaille; Bruce Drinkwater (professor of ultrasonics, University of Bristol);

Maria Geffen (associate professor of otorhinolaryngology, neuroscience, and neurology, University of Pennsylvania School of Medicine); Dame Evelyn Glennie; Nobuhiro Hagura (senior researcher, National Institute of Information and Communications Technology); David Hopkin; Tom Hunt; Gil Menda (Department of Neurobiology and Behavior, Cornell University); Dom Monks; Paul Musgrove; Will Musgrove; Garrod Musto; Jim Reekes; Steve Taylor; Luca Turin (professor of physiology, University of Buckingham); Suzie Yuan; Joshua Zeman.

Many would consider it folly to endeavor to write a pop science book with only scant scientific knowledge. After all this time, I might now agree. However, at the outset, I sensibly turned to three fantastic "science friends" to check my facts and help fill the gaping holes in my scientific knowledge. Their time, energy, patience, and knowledge have been invaluable. My sincerest thanks to Mark Grinsell, Joe Sidders, and Patrick Slade.

When I began to pitch the concept of *What the Ear Hears (and Doesn't)* to literary agents, I didn't seriously believe that any of them would be particularly interested—after all, who in their right mind would trust a musician to write a pop science book with the pretty bonkers concept of an Infinite Piano as its narrative? Well, Donald Winchester at Watson Little did. I will be forever indebted to him for placing his faith in me, a first-time author with an idea and a couple of very badly written sample chapters. I've never worked with another literary agent, but if they're all as supportive, encouraging, patient, and skilled as Donald, it must be a world-leading profession!

And my eternal thanks go also to the fantastic team at Sourcebooks: to Brittney Mmutle; to Jillian Rahn and Stephanie Rocha for their inspired art/design work; to Philip Pascuzzo for the beautiful cover design; to Tara Jaggers for turning my very sketchy sketches into wonderful illustrations; to production editor Emily Proano and proofreader Carolyn Lesnick for their incredible attention-to-detail; and my editor, Erin McClary, who has been a fantastic support, encouraging and enthusiastic at all times. Her innate sense of narrative and structure has been a brilliant guide for me, helping me shape the book into a work far better than I could ever have imagined.

Finally, to my wonderful family, Juliette, Rose, and Jack, the center of my world—you can come in now. I've finished it!

INDEX

Note: Page numbers in *italic* refer to illustrations.

A

abiotic sounds, 61

air-raid sirens, 197–199, *199*

alarm tones, 200–202

Also sprach Zarathustra (Strauss), 19–20

Angliss, Sarah, 92–93

animals, 9–10, 57–68, 121–142, 146, 155, 186–189

anthropogenic noise, 126–129, 135

antinodes, 160, *161*, 238–239

Apollo missions, 28–29

Apple's start-up chime, 34–38, *35*

atomic clocks, 261–263

atomic weapon tests, 249–254

aurorae, 29–31, *31*

autofocus, 208–212, *211*

Auto-Tune, 215–221, *217*

B

babies, 147–149, 151, 152–153

Bach, J. S., 93–94, 154

Bagno, Samuel, 206–208

banana equivalent dose (BED), 10

bass instruments, 85–88, *86*, 111

bees, 135–138

Beethoven, Ludwig van, 45, *46*, 98–99

"Believe" (Cher), 217–218

Bell Labs, 23–25

Berlioz, Hector, 86–87

"Bikini snow," 249–254

birds, 62–68, *65*, 128

black hole, in the Perseus galaxy
cluster, 13, 17–19, *18*

blending frequencies, 98–120,
197–199

body, resonant frequencies of, as
chord, 83, *84*

Bose, Amar, 116–117

brain, human, 50, 90, 147–161,
230–232, 234–236

bridges, 68–76

Broughton Bridge collapse,
68–69

bubbles, 95–97

Buys Ballot, Christophorus
Henricus Diedericus, 203–206

C

cameras, autofocus, 208–212, *211*

Camster Round (Neolithic
grave), 88–89

car parking sensors, 159–161, *161*

cats' purrs, 9–10

caves and prehistoric tombs,
88–91

Chandra X-ray Observatory, 13,
17–19, *18*

chanting, 90–91

Charles, Prince of Wales, 144

Cher, 217–218

Chernobyl, 255

cherry analogy, 79–80, *80*

children, 147–149, 151, 152–153,
189–192, 202

Chopin, Frédéric, 111–112

Choral no. 3 (Franck), 94

Chun Quoit (prehistoric tomb),
89–90

classical music, 85, *86*, 154–155

clocks and watches, 258–264,
260

cochlea, 183–184, *185*

cochlear implants, 182–183

Cole, USS, attack on, 199–200

color, 162–169

compression, 115

Concorde (airliner), 63, 66–68,
66

Creation, The (Haydn), 21

cucumber analogy, 214–215

D

dampers, 75

Death and the Maiden (Schubert),
134

deep sound channel, 122, *122*

Delibes, Léo, 231–232

Deutsch, Diana, 225–227, 233, 234

Devereux, Paul, 89–90

devil's interval, 100

difference tones, 106–109, 199

Different Trains (Reich), 234–235

disasters, 52–76

divers, 95–97

Doppler shift, 66, 203–206, *205*, 207

Drinkwater, Bruce, 158–161

duck analogy, 203–204

Dukas, Paul, 87–88

Dunkirk (film), 221–223

E

earthquakes, 55–62, 64

Earth, resonant frequency of, 25–27, *26*, 50–51

Earth's orbital velocities, 14–15, *15*

echocardiography, 244

electric pianos, 41–45, *43*

electric power lines, 135

electromagnetic spectrum, 163–164, 247–249, *248*

electromagnetic waves, 8, 163

elephants, 57–62

ELF EMF (extremely low frequency electromagnetic fields), 135

endurance flight, 116–117

equalization (EQ), 112–115

equal temperament, 102, 105–106

Eroica (Beethoven), 98–99

esters, 179

extremely low frequency electromagnetic fields (ELF EMF), 135

eyeballs, human, 81, 97

F

Fender Rhodes pianos, 43–45, *43*

fifth intervals, 99

52-hertz whale (the loneliest whale), 123–126, 146

film music, 221–223

fire extinguishers, sonic, 117–118

Fletcher, Harvey, 201–202

flight, endurance, 116–117

"forkestra," 40–41, *40*

fragrances, 174–179

Franck, César, 94

frequency, and the "Infinite Piano," 5–8

frequency, impact of, on humans, 8–9

frequency, related to wavelength, 167–168, *168*

frequency, unit of (Hz), 7

G

Gagliano, Monica, 145–146

Geffen, Maria, 139–142

genitalia, 172–174

"George Michael reverb," 115

Gershwin, George, 195–196

"ghostly" encounter, 77–81, *80*, 84–85, 88

Gideon, 194

Glennie, Evelyn, 180–181

glissando, 195–197, 222–223

Godzilla, 252

Gorman, Ross, 195–196

Grieg, Edvard, 87–88

Guido d'Arezzo, 155–156

H

Hagura, Nobuhiro, 169–172

Handel's tuning fork, 39–40, 45, *46*

Haydn, Joseph, 21

Hayek, Nicolas, 258, 260

headphones, noise-canceling, 115–117

hearing, sense of, 57, 179–185, *185*, 188–189, 201–202

Helmholtz, Hermann von, 166–167

hertz (Hz), definition of, 7

Hildebrand, Andy, 215–221

Holst, Gustav, 20–21

honeybees, 135–138

Hooke, Robert, 46–49, *48*

horse racing fraud incident, 186–189

House, William F., 182–183

Hudson, Joseph and James, 107–108

human body, resonant frequencies of, as chord, 83, *84*

Hz (hertz), definition of, 7

I

Infinite Piano, 5–8

"infinite" scales, 221–225

infrasound, 20, 57–68, *65*, 77–97

InSight mission, 27–28

International Organization for Standardization (ISO), 34, *46*, 49

interstellar space, 23

intervals, 14–16, *14*, 99–101

"In the Hall of the Mountain King" (Grieg), 87–88

ionosphere, 25–27, *26*, 30

ISO (International Organization for Standardization), 34, *46*, 49

J

Jansky, Karl, 23–25, *24*

Jericho's trumpets, 192–195

John Somebody (Johnson), 235

Johnson, Scott, 235

Jonathan Hagstrum, 62–63, 66–68

Jones, Reginald Victor, 2–5

K

Kepler, Johannes, 13–16

Knickebein project, 1–5, *2*

Krakatoa eruption, 158, 241

Kraus, John, 25

L

Lakmé (Delibes), 231–232

Laming, James, 187–189

La nativité du seigneur (*The Birth of the Lord*; Messiaen), 94–95

languages, tonal, 232–233

levitation, 158–161, *161*

light, 8, 162–169, *163*

lightning flashes, 25–27

Lombard effect, 127–128

"loneliest whale" (52-hertz whale), 123–126, 146

Long-Range Acoustic Device (LRAD), 199–202

Lorenz system, 1–5, *2*

loudness, perception of, 201–202, *202*

lowest "note" in the universe, 17–19, *18*

low-frequency sound. *See* infrasound

LRAD (Long-Range Acoustic Device), 199–202

Lucky Dragon 5 incident, 249–254

lullabies, 147–149

M

Mabaan tribe, 182

Mac computer start-up chime, 34–38, *35*

"magic" powers, 49–51, 156–159

major third, 99–100

Mandarin Chinese, 232–233, *233*

Mars, 27–28, 264

Marshall Islands, 249–254

Maxwell, James Clerk, 164–165, 168–169

mechanical resonance, 54

mechanical waves, 8

Menda, Gil, 130–132

Messiaen, Olivier, 94–95

Michael, George, 114–115

microseisms, 63, 64–66, 65

microwave ovens, 237–239, 239

Milky Way radio emissions, discovery of, 23–25, 24

Millennium Bridge footbridge, 72–76

Minnaert resonance, 96

molecular vibrations, 176

mondegreens, 230–232

Monks, Dom, 118–120

moon, 28–29

Mosquito device, 189–192

motion sensors, 206–208

"Mozart effect," 152–155

Munson, Wilden A., 201–202

musical frequency table, 295–297

musical intervals, 14–16, 14, 99–101

musical notation system, 155–156

musical standardization, 32–39, 45–51, 46

musical tuning systems, 101–106, 104

Music of the Spheres, The (Mainwaring), 16

music, space, 16, 19–23

Mussorgsky, Modest, 87

N

NASA, 81–83

Navy, U.S., 121–122, 122, 127, 199–200

"near-field" hearing, 129

Newton, Isaac, 165–166

Night on the Bare Mountain (Mussorgsky), 87

Nocturne in E Minor (Chopin), 111–112

nodes, 160, 161, 238–239

noise-cancelling headphones, 115–117

nuclear tests, 249–254

O

obstetric ultrasound, 242–243

ocean, sounds of, 63, 95–97, 121–128, 122

olfaction, 174–179

organs, pipe, 92–95

overtones, 90–91, 109–111, 110

P

Pathétique (Tchaikovsky), 223–227, *226*

Penderecki, Krzysztof, 196–197, *197*

perfect pitch, 49, 233

perfumery, art of, 177–179, *178*

Perseus galaxy cluster's music note, 13, 17–19, *18*

pianos, electric, 41–45, *43*

Piano, Infinite, 5–8

piano tuning, 101–102, 106–107

pigeon disappearance mystery, 62–68, *65*

pipe organs, 92–95

pitch correction, 213–221, *217*

pitch, scientific, 49–51

pitch, standardization of, 32–39, 49

planets' orbital velocities, 13–16, *15*

Planets, The (Holst), 20–21

plants, 10, 143–146, *145*

plosives, 113–114

Pogo oscillation, 82–83

Polaroid's autofocus camera, 208–212, *211*

police whistles, 107–109

pop shield (or filter), 113–114

power lines, 135

prehistoric tombs and caves, 88–91

presbycusis, 181–182, *185*, 188–189

printers, inkjet, 245–247, *246*

Puleo, Joseph, 156–157

pyramids of Egypt, 158–159

Pythagoras, 13

Pythagorean Tuning, 102–106, *104*

Q

quantum biology, 177

quartz, use in timekeeping, 258–261, *260*

R

radiation, 10, 247–257, *248*, *255*

radio beams, used in WWII bombing raids, 1–5, *2*

radio emissions from Milky Way, discovery of, 23–25, *24*

rainbows, 162, *163*, 164, 165–166

rarefaction, 115

rats' whiskers, 138–142, *140*

Reber, Grote, 25

recipe analogy, 91

Reekes, Jim, 35–38

Reich, Steve, 234–235

religious feelings, 92–95

resonance/resonant frequencies, 52–55, 68–76, 83, *84*, 88–91, 96, 138–142, 176, 262. *See also* Schumann resonance

Rhapsody in Blue (Gershwin), 195–196

Rhodes pianos, 41–45, *43*

rhythm, use in piano tuning, 101–102

Risset, Jean-Claude, 222–223

Robertson, Seth, 117–118

Romantic music, 85–88, *86*

Röntgen, Wilhelm, 17, 256–257

Rutan, Dick, 116–117

S

Sagan, Carl, 22

Savart, Félix, 48–49

Schubert, Franz, 134

Schumann resonance, 25–27, *26*, 50–51

scientific pitch, 49–51

Seabourn Spirit ship, pirate attack on, 200

Seiko, 258–259

seismic activity, 28–29, 55–68, *65*

seismometers, 64

semitones, 100–101

senses, human, 162–185

sexual function, 172–174

She Goes Back under Water (Angliss), 92–93

Shepard, Alan B., 182–183

Shepard, Roger (Shepard tone), 221–223

shofars (rams' horns), 192–195

Shore, John, 38–39

sibilants, 114–115

singing, 88–91, 112–115, 147–149, 213–221, *217*

sirens, 197–199, *199*

"sleights of ear," 222–227, *226*

smear. *See* glissando

smell, sense of, 174–179

SOFAR (sound fixing and ranging), 122, *122*

solar wind, 22–23

solfeggio frequencies, 156–157

"sometimes behave so strangely" phrase, 234

sonar, 126–127, 209

sonic booms, 66–68, *66*

Sorcerer's Apprentice, The (Dukas), 87–88

SOSUS (Sound Surveillance System), 121–122, *122*, *123*

sound, as mechanical wave, 8

sound barrier, breaking, 67

sound fixing and ranging
(SOFAR), 122, *122*
Sound Surveillance System
(SOSUS), 121–122, *122*, 123
spaceflight/space missions, 22–
23, 27–29, 81–83, 264
space music, 16, 19–23
speech-to-song illusion,
234–235
Spencer, Percy, 237–239
spiders, 128–135
spiderwebs, 133
spiritual experiences, 91, 92–95
standing waves, 79–81, *80*, 88,
238
Stapleton, Howard, 189–192
Strauss, Richard, 19–20
string sections, orchestral,
227–230
studio recording, 112–115, 118–
120, 213–221
Stuka dive bombers, 194–195
stun gun conspiracy, 186–189
surfing analogy, 167–168, *168*
Sutton, Thomas, 165, 168–169
"sweet spots," 54, 89
Symphonie fantastique (Berlioz),
86–87
Szechuan peppers, 169–172

T

Tacoma Narrows Bridge
collapse, 70–72
"Tae Bo Twenty-Three," 52–55,
53
Taipei 101 skyscraper, 75
Tandy, Vic, 77–81, 84–85, 88
tarantula-inspired words,
133–134
Tartini's undertones, 106–109,
199
taste, sense of, 169–172
Tchaikovsky, Pyotr Ilyich, 223–
227, *226*
Techno Mart tower vibrations,
52–55, *53*
teenagers, 189–192, 202
Tesla, Nikola, 55–57
*Threnody for the Victims of
Hiroshima* (Penderecki), 196–
197, *197*
time, 258–265
Toccata and Fugue in D Minor
(Bach), 93–94
tonal languages, 232–233
toothbrushes, 239–242
touch, sense of, 172–174
trains, 203–206, 263–264
Tran, Viet, 117–118
tritones, 100
tsunamis, 57–62

tuning forks, 38–41, 43, 45–46, 46

tuning, standardization of, 45–49, 46

tuning systems, 101–106, 104

Turin, Luca, 175–179

Tuvan throat singing, 90–91

U

ultrasound, 159–161, 161, 186–192, 206–212, 211, 239–244, 243, 259

undertones, 106–109, 199

underwater sounds. See ocean, sounds of

U.S. Navy, 121–122, 122, 127, 199–200

V

vibrato, 228–229

vibrators, 173–174

vision, color, theory of, 166–167

voicing chords, 111–112

Voyager missions, 22–23

W

"waggle dance," 135–138

wailing sounds, 195–199

watches and clocks, 258–264, 260

Watkins, William, 123–125

wavelengths, 144, 167–168, 168

waves, concept/types of, 7–8

whales, 123–128, 146

whiskers, rats', 138–142, 140

whistles, two-tone, 107–109

whole body hearing, 179–181

Woods Hole Oceanographic Institution, 123–125

X

X-rays, 17, 256–257

Y

Yeager, Jeana, 116–117

Young, Thomas, 166–167

ABOUT THE AUTHOR

Richard Mainwaring is a performing musician, composer, TV presenter, and educator. He has composed music for Netflix, the BBC, and the International Space Station. He is a classically trained multi-instrumentalist and has worked with artists as diverse as Hugh Laurie, Hayley Westenra, Larry Adler, Eartha Kitt, and Tom Odell. He has also presented over fifty short films for the BBC.